Ecotoxicology

Effects of Pollutants on the Natural Environment

Ecotoxicology

Effects of Pollutants on the Natural Environment

Colin Walker

CRC Press is an imprint of the
Taylor & Francis Group, an **informa** business

CRC Press
Taylor & Francis Group
6000 Broken Sound Parkway NW, Suite 300
Boca Raton, FL 33487-2742

Printed on acid-free paper
Version Date: 20131217

International Standard Book Number-13: 978-1-4665-9179-0 (Paperback)

Library of Congress Cataloging-in-Publication Data

Walker, C. H. (Colin Harold), 1936- author.
Ecotoxicology : effects of pollutants on the natural environment / Colin Walker.
pages cm
Summary: "During the latter part of the 20th century chemical industry grew rapidly, and with this growth new industrial chemicals found their way into the natural environment. Pesticides came to be used on a larger scale, and questions were asked about residues of them that were found in environmental samples (biota, soil, water, and air). Residues were also found of a range of industrial chemicals in effluents entering rivers and polluted air. Some of these events received extensive media coverage, which was something of a mixed blessing. While important discoveries were made known to a wide audience, inaccuracies and half-truths crept into this reportage, sometimes leaving a rather confusing impression to interested laypeople. In time, government laboratories, industrial laboratories, and universities became involved in the investigation of pollution problems, and the discipline of ecotoxicology began to take shape. Today ecotoxicology courses are offered by universities and colleges of further education. While a number of textbooks are now available to students who follow ecotoxicology courses at universities and other institutions of higher education, there appears to be a shortage of texts aimed at interested laypeople. This seems unfortunate, because the science underlying environmental pollution has intriguing aspects to it. There is much evidence for the phenomenon of chemical warfare in nature, which, over a long period, has been a driving force in the evolution of plant toxins and the production by animals of systems that detoxify them. The selective pressure of pesticides has led to the evolution of resistant strains of pests. The biomagnification of recalcitrant organic pollutants in food chains has raised problems for predators of higher"-- Provided by publisher.
Includes bibliographical references and index.
ISBN 978-1-4665-9179-0 (paperback : acid-free paper)
1. Environmental toxicology. 2. Ecological risk assessment. 3. Environmental chemistry. I. Title.

QH545.A1W35 2014
577.27--dc23 2013048047

Visit the Taylor & Francis Web site at
http://www.taylorandfrancis.com

and the CRC Press Web site at
http://www.crcpress.com

Contents

Preface

During the latter part of the twentieth century the chemical industry grew rapidly, and with this growth new industrial chemicals found their way into the natural environment. Pesticides came to be used on a larger scale, and questions were asked about residues of those that were found in environmental samples (biota, soil, water, and air). Residues of a range of industrial chemicals were also found in effluents entering rivers and polluted air.

Some of these events received extensive media coverage, which was something of a mixed blessing. While important discoveries were made known to a wide audience, inaccuracies and half-truths crept into this reportage, sometimes leaving a rather confusing impression to interested laypeople.

In time, government laboratories, industrial laboratories, and universities became involved in the investigation of pollution problems, and the discipline of ecotoxicology began to take shape. Today ecotoxicology courses are offered by universities and colleges of further education.

While a number of textbooks are now available to students who follow ecotoxicology courses at universities and other institutions of higher education, there appears to be a shortage of texts aimed at interested laypeople. This seems unfortunate, because the science underlying environmental pollution has intriguing aspects. There is much evidence for the phenomenon of chemical warfare in nature, which, over a long period, has been a driving force in the evolution of plant toxins and the production by animals of systems that detoxify them. The selective pressure of pesticides has led to the evolution of resistant strains of pests. The biomagnification of recalcitrant organic pollutants in food chains has raised problems for predators of higher trophic levels. Sublethal effects of pollutants have caused populations to decline.

The present book has grown out of considerations such as these. The intention has been to write an account of ecotoxicology that highlights these issues without going into the detail of a standard textbook. A glossary has been included to explain terms that may be unfamiliar to some readers.

The author is grateful to Richard Sibly, Robin Hewison, Bob Olliver, Bob Lugg, Neil Walker, and Joan Millard, all of whom have given valuable advice and suggestions during the writing of this book.

Introduction

This book is about the effects of chemicals upon the living environment. It is a subject that first gained the attention, interest, and concern of a wide audience during the second half of the twentieth century when a number of serious pollution problems was given extensive coverage in the media. These were, very largely, the outcome of human activity; for example, the discovery of new pesticides, the release of heavy metals during mining operations, and the release of pollutants with sewage effluents into surface waters. Rachel Carson's widely read book *Silent Spring*, published in 1963, focused on the harmful effects of pesticides. While not always scientifically accurate, it drew attention to a serious environmental issue and paved the way for the introduction of more stringent legislation to control the marketing of pesticides and other biocides in Western countries.

So, during the latter part of the last century, there came recognition of pollution problems such as these that were the consequences of the activities of man. Pollution caused by man was known to go back to the early history of urbanization when smoke from fires and furnaces and the release of untreated sewage caused pollution problems in towns. At the time, the concern was about harmful effects upon humans. It was not until the second half of the twentieth century that serious consideration was given to toxic effects of chemicals on the natural environment.

Looking back into the past, toxic effects of chemicals upon the natural environment have been occurring over a long period. Natural events such as volcanic eruptions, earthquakes, forest fires, and the landing of meteorites can all cause the release of abnormally high levels of metals such as copper or lead and gases such as sulfur dioxide and nitrogen oxides, all of which are potentially toxic to living organisms. Thus, it is reasonable to suppose that environmental chemicals have been having toxic effects upon living organisms since very early in the history of life on earth. Indeed, parallels can be drawn between (1) environmental pollution caused by mining and the smelting of ores and (2) toxic effects by the same compounds as a consequence of extreme natural events such as volcanic eruptions.

Chemical warfare is commonplace in nature, where certain species use chemical weapons against others, to their own evolutionary advantage (see, for example, Ehrlich and Raven 1964; Harborne 1993). Plants synthesize toxins that are harmful to the animals that feed upon them, thus giving them some protection against grazing. Animals also possess chemical weapons for both attack and defense. Thus, some predators (e.g., many spiders, scorpions, and snakes) produce venoms that poison and paralyze their prey, and some prey species produce chemicals that are toxic to species that prey upon them. In the second category, the bombardier beetle has a formidable weapon. It can fire a hot solution of irritant quinones at its assailants. Further, herbivorous animals have been found to possess enzymes that can break down toxins present in the vegetation upon which they feed. Following from this, it is suggested that there has been a coevolutionary arms race leading to the emergence of such protective detoxifying enzymes in animals (see Harborne 1993; Lewis 1996).

Chemical warfare has had a long evolutionary history. The discovery and development of chemical weapons by man is a very recent event on this scale.

Among the earliest known examples of chemical weapons employed by humans are arrow poisons, still used by some indigenous peoples in areas of Africa, Latin America, Asia, and Oceania. Such weapons have been used for both warfare between tribes and hunting. One example of an arrow poison, curare, has been employed by Amazonian Indians. It is extracted from certain tree resins. Interestingly, it has found a use in human medicine; at carefully controlled doses it acts as a muscle relaxant! Another example is the cardiac glycoside ouabain. Plant extracts containing ouabain are used in some parts of Africa as arrow poisons. In the well-known Sherlock Holmes story *The Sign of Four*, an Andaman Islander uses a blowpipe to deliver poison darts to deadly effect.

In recent times, some natural products have been used for pest control, in the chemical war waged by humans against pests, weeds, disease vectors, and other organisms not seen as beneficial to the human race. Nicotine, pyrethrum, and rotenone have all been used as insecticides. Strychnine has been used to poison moles. Most importantly, some of these compounds have provided models for the development of new pesticides by the chemical industry. They have revealed mechanisms of toxicity that can be exploited for the purposes of pest control. Thus, the pyrethroid insecticides have been based on the structure of constituents of pyrethrum (pyrethrins), and the neonicotinoids have been modeled on the structure of nicotine. In a sense, evolutionary history is repeating itself, except that humans instead of natural selection are now determining the course of events. Many drugs have also been developed from natural products. Natural products are a rich source of biologically active compounds in general.

When new pesticides are released into the environment, they are detoxified by animals, largely due to their metabolism by enzyme systems, many of which are believed to have evolved to give protection against plant toxins. Not altogether surprising in the case of pesticides that have been modeled upon natural products. Thus, the ecotoxicology of pollutants should be seen against the background of the long history of chemical warfare in nature, a conflict that is still in evidence today.

Returning to developments during the latter part of the twentieth century, as more evidence came to light of pesticides and other man-made chemicals having harmful effects upon the living environment, the discipline of ecotoxicology began to take shape. From a number of well-documented case studies, some basic processes and principles began to emerge. Persistent organohalogen compounds such as the organochlorine insecticides dieldrin and DDT and certain polychlorinated biphenyls (PCBs) were found to undergo biomagnification in food chains. Such biomagnification sometimes led to toxic effects upon predatory species (e.g., raptors) at the top of food chains, because of the high levels of organochlorine compounds in their prey. The molluscicide tributyl tin was shown to cause endocrine disruption in certain molluscs (e.g., the dog whelk). A component of the contraceptive pill (EE2) was found to cause feminization, and consequent infertility, in male fish below sewage outfalls. Recently it has been shown that some neurotoxic insecticides can cause behavioral disturbances in bees. Many of these effects have been related to population declines.

Another aspect of ecotoxicology has received particular attention because of its economic importance—the development of resistance. With the overuse of pesticides, resistant strains of pest species, vectors of disease, and weeds have emerged—following Darwinian principles. Unnatural selection has operated in favor of genes that confer resistance, with consequent loss of pest control. Biochemical studies have uncovered molecular mechanisms responsible for the development of resistance in major pest species. Such research has been of considerable interest to both manufacturers of pesticides and farmers. For example, the discovery of the mechanism by which an insect pest develops resistance can guide the way to discovering a new related insecticide that is not affected by this mechanism—and which, consequently, may give effective control of resistant strains of the pest. Studies of this kind are illustrative of certain evolutionary processes that are in operation today, and provide interesting material for evolutionary biologists.

In writing the present text, the aim has been to explain some basic principles of ecotoxicology, and to clarify the science behind chemical pollution of the natural environment and the strategies for pollution control. Considerable use has been made of case studies that illustrate these principles. The text has been divided into three sections. The first, "Basic Issues," outlines these principles, making reference to illustrative examples given later in the text. The second part focuses on particular types of pollutants and describes relevant case studies. The third and concluding section contains three chapters dealing with wider issues and attempting to look into the future.

Technical terms have been used sparingly, and a glossary has been supplied to explain terms that may be unfamiliar to readers.

Author

Colin Walker originally qualified as an agricultural chemist. He was responsible for chemical and biochemical studies of environmental pollutants at the Monk's Wood Experimental Station during the mid-1960s when certain effects of organochlorine insecticides were established. This work led to restrictions on the use of cyclodienes and DDT. He subsequently joined the University of Reading where he taught and conducted research on the molecular basis of toxicity with particular reference to ecotoxicology. Now retired, he is currently affiliated with the Department of Biosciences at the University of Exeter where he contributes to the teaching of a course in ecotoxicology.

Section I

Basic Issues

This first section of the text will be concerned with a definition of terms and explanation of some basic principles of ecotoxicology. Included here will be an account of chemical warfare in nature and its evolutionary history, as well as the hazards presented to the natural environment by chemicals released by humans into the environment. The mechanisms of toxicity are explained, and the question of sequential effects of pollutants at different organizational levels addressed. Attention will be given to the particular problems presented by pesticides and the development of resistance to them. Naturally occurring pollutants and the effect of chemicals on natural processes will be described, leading to a consideration of global models that incorporate phenomena such as global warming and damage to the ozone layer.

Section 2 will give examples of pollution that illustrate principles explained in this first section.

1 Toxicology and Ecotoxicology

SOME DEFINITIONS

Toxicology is concerned with the harmful effects of chemicals upon living organisms. Chemicals that have harmful effects are described as poisons, a term that requires careful definition. This issue was first addressed by Paracelsus (1493), who famously said (in free translation): "All substances are poisons and there is none that is not a poison. It is the dose that makes the poison." In other words, everything can be poisonous if the dose is high enough, but if the dose is sufficiently low, nothing is. This remains essentially true, even though it may be very difficult to administer a dose of some chemicals that is high enough to be harmful. It does, however, argue for caution when using the word *poison*. A substance can only be poisonous when it is given above a certain dose. At lower doses it is not poisonous. This simple principle is often not recognized in popular articles on pollution. Sometimes poisons are reported to exist in food or water at concentrations so low that there is no known risk to the living organisms that encounter them. Thus, the public can be alarmed through disinformation. A little sensation does help to sell newspapers.

There is also the question: What constitutes harm? Simple enough to understand in the case of toxicity tests that use lethality as the endpoint. Here toxicity can be quantified as a median lethal dose (LD_{50}) or median lethal concentration (LC_{50}), for example. But there are many other measures of harm apart from lethality, e.g., neurotoxicity, carcinogenicity, endocrine disruption, etc. Included here are biochemical effects (e.g., inhibition of brain acetylcholinesterase or DNA damage), physiological effects (e.g., disturbance of nerve function), behavioral effects (e.g., responses of animals to stimuli), etc. In human toxicology many indices of harm other than lethality are used in toxicity testing.

The term *ecotoxicology*, first introduced by Truhaut in 1969 (Truhaut 1977), suggests an area of science that brings together ecology and toxicology. It is here defined as "the study of the harmful effects of chemicals upon ecosystems." Ultimately, the greatest concern is about effects at the levels of population, community, and ecosystem. However, an essential part of the subject is the identification of harmful effects at the individual level, which may be translated into these higher-level effects in the field. Important here are the toxic effects of pesticides upon individual pests that lead to the development of resistance at the population level. Ecotoxicity tests that can identify such lower-level effects (e.g., mechanistic biomarkers) have the potential to give forewarning of longer-term change at the population level. Such effects may become evident in field studies, including field trials that are carried out with pesticides or other hazardous environmental chemicals. Mechanistic biomarker tests may be employed in the early stages of environmental risk assessment (Chapters 2 and 19).

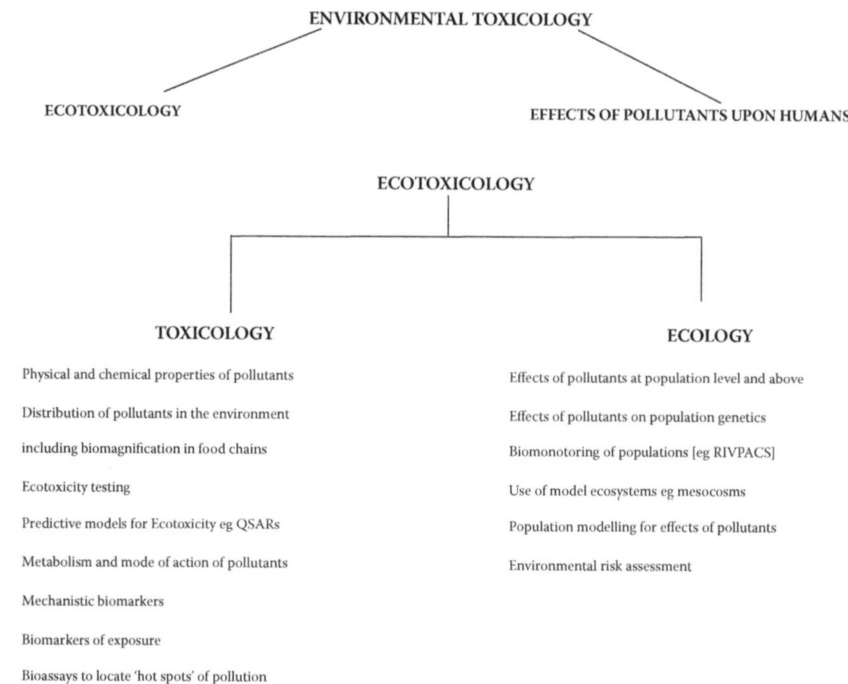

FIGURE 1.1 Relationship of ecotoxicology to other disciplines.

Returning to questions of definition, in human toxicology the effects of chemicals upon individuals are of primary concern, but in ecotoxicology the focus is on effects at the level of population and above. Studying the exposure of humans to hazardous chemicals in the environment does not fall within the definition of ecotoxicology given above and will not be dealt with in this book. Thus, the effects of pesticides upon those who use them in the field, e.g., agricultural workers or pilots involved in aerial spraying, will be regarded as an aspect of human toxicology. The broader term *environmental toxicology* will be taken to include both ecotoxicology and questions about human health hazards arising from exposure to chemicals in the environment (Figure 1.1).

THE DISCIPLINE OF ECOTOXICOLOGY

Aspects of ecology and toxicology that are of importance in ecotoxicology are identified in Figure 1.1. To take toxicological aspects first, the physical and chemical properties of pollutants are critical determinants of their movement, distribution, and persistence. Biochemical and physiological properties (e.g., toxicity, mode of action, and metabolism) determine the effects that pollutants have on living organisms. Metabolism is also critical in determining the fate of pollutants in living organisms and, consequently, whether they undergo biomagnification with movement through food webs. Compounds that are readily biodegradable tend not to undergo biomagnification.

The development of appropriate tests for ecotoxicity is a particularly difficult area. Questions are frequently asked about toxic effects on species that cannot realistically be tested in the laboratory—for reasons of rarity, cost, protection, etc. As a rule, surrogate species are used. These issues will be discussed further in Chapters 2 and 8. Related to this, there is an interest in the development of predictive models for ecotoxicity, which can, in theory, provide a cost-effective alternative to ecotoxicity testing per se (Chapter 19).

An important developmental area is the design of mechanistic biomarker assays. The aim here is to develop user-friendly assays that can provide measures of toxic effect on living organisms in the field after exposure to pollutants (Chapters 2 and 19). Related to this, bioassays have been developed that can be used to identify hot spots of pollution. Included here are cell lines, such as the Calux system, which can detect low levels of coplanar polychlorinated biphenyls (PCBs) and dioxins that interact with the Ah receptor (see Chapter 13).

Turning to ecological aspects, effects at the population level are of primary importance. Population studies in the field and population modeling are relevant here (Chapters 4 and 18). The use of mesocosms and other model ecosystems is an interesting developmental area (Chapters 2 and 4). The use of mechanistic biomarkers, mentioned earlier, can provide vital support for population studies in the field. Effects on the genetic composition of populations are also of great interest, as in the case of development of resistance to pesticides (Chapter 5). Other important ecological topics are biomonitoring for effects of pollutants and environmental risk assessment.

The terms *contaminant* and *pollutant* are used inconsistently in the literature. Throughout this text the broad term *environmental chemical* will be applied to any substance that is found in the environment, without any implication about its abundance or the hazards that it may present. On the other hand, the term *contaminant* will be applied very largely to chemicals that are generated by man and do not normally occur in nature. It will occasionally be applied also to certain naturally occurring substances, when they are found at abnormally high levels, e.g., heavy metals in the vicinity of mine workings or SO_2 gas when it is released at the time of a volcanic eruption. Such environmental chemicals present the same problems to the environment whether they originate from natural or human activity. When detected in the environment they will often originate from both natural and unnatural sources, which are difficult or impossible to distinguish between. It seems logical to simply refer to them as being contaminants regardless of the extent to which they are natural or unnatural in origin, the important point being that they occur at abnormally high levels so far as ecosystems are concerned.

The term *contaminant* does not necessarily indicate that a chemical is hazardous to the environment. The term *pollutant*, however, will be reserved for contaminants for which there is clear evidence of their capacity to cause harm at environmentally realistic concentrations. As a text in ecotoxicology, this book will focus upon situations where contaminants are at sufficiently high levels to cause harm to ecosystems, and may therefore be regarded as pollutants. Chemicals referred to as pollutants in the following text can occur at high enough concentrations to be termed poisons according to the definition of Paracelsus.

The pollution of the environment described here is predominantly the consequence of the activities of man. This is largely self-evident, because the chemicals in question are only produced by man. However, we come back to the question: When can naturally occurring chemicals be regarded as pollutants? Most chemicals that answer to this description are inorganic. They include metals such as cadmium, lead, zinc, copper, nickel, manganese, and mercury and the gases sulfur dioxide, hydrogen sulfide, and nitrogen oxides. Usually these only reach levels that can cause environmental harm because of the activities of man. Thus, lead, cadmium, and mercury have caused pollution problems as a consequence of mining or the activities of the chemical industry. Pollution by sulfur dioxide, which has caused extensive damage to forests in parts of Eastern Europe, has resulted from the burning of brown coal. Further examples will be given in the following text. However, there are, as usual, exceptions to the rule. Pollution of this kind may also arise as the consequence of cataclysmic natural events, such as earth tremors or volcanic action; it is not always due to the activities of man. Disturbances of this kind may expose minerals containing metals to weathering, which can lead to a substantial rise of the concentrations of them in surface waters and soils. Also, volcanic eruptions can cause the release of the gases sulfur dioxide, hydrogen sulfide, and nitrogen dioxide at high enough levels to cause damage to animals and plants. The destructive power of volcanoes was well illustrated—and well recorded—when Mount Vesuvius erupted in 79 AD. The description of this cataclysmic event by the Roman historian Pliny provides the basis for Robert Harris's novel *Pompei* (2003).

Most examples of pollution by naturally occurring substances have been caused by inorganic chemicals, but there are also a few naturally occurring organic chemicals that have been implicated. For example, some heavy metals/metalloids are converted by microorganisms into methyl derivatives. Examples include mercury and arsenic. As will be discussed later, organometallic compounds have different properties from the metals/metalloids from which they are derived—and often they are more hazardous to living organisms. Some of them have been implicated in serious pollution problems. At Minamata Bay in Japan, between the late 1950s and the early 1960s, high levels of methylmercury in the marine environment had toxic effects on fish—and on human beings feeding on the fish. At that time methylmercury was used extensively as a fungicide; environmental residues of methylmercury arose from both release of the manufactured organic form and environmental synthesis from inorganic mercury (Environmental Health Criteria 86 and 101 (WHO 1989)).

Fires due to natural phenomena such as volcanic eruptions or lightning strikes can also lead to pollution. When trees or dry vegetation burn, a diversity of organic compounds is formed. These include products of incomplete combustion, such as polycyclic aromatic hydrocarbons. Among these are some potent carcinogens such as benzo(a)pyrene—one of the compounds in cigarette smoke that is associated with the development of lung cancer by smokers. Thus, with extensive bush or forest fires certain natural organic pollutants may be spread over large areas. Such pollution of the natural world will have occurred long before the appearance of man on the planet.

In all of these examples, whether the consequence of human activity or of natural phenomena, it is likely that the pollutants will have exerted some selective pressure on the populations that are exposed to them. This issue will be discussed further in Chapter 5.

SELECTIVE TOXICITY

As we have seen, toxicity is dependent on dose, and this means that the term *poison* needs to be used with caution. In reality, a chemical is only poisonous if the dose is sufficiently high. It is possible, however, to distinguish between chemicals that have low, medium, high, or very high toxicity to particular organisms, the distinction being based on how high a dose is required to produce a defined toxic effect. This approach has been used in the text *Basic Guide to Pesticides: Their Characteristics and Hazards* by S. Briggs (1992). Here, for example, the following categories are defined for acute (immediate) oral toxicity to animals:

Toxicity Rating	Dose Range
Very high	<50 mg/kg
High	50–500 mg/kg
Medium	500–5000 mg/kg
Low	>5000 mg/kg

These figures illustrate the huge range of toxicity values found for different pesticides across a range of species. It should be remembered when looking at these figures: the higher the lethal dose, the lower the toxicity.

Any one chemical is often much more toxic to some species than it is to others; in other words, many chemicals show marked *selective toxicity* between different species, sexes, strains, or age groups. For example, most herbicides are much more toxic to plants than they are to animals—and most insecticides are much more toxic to insects than they are to plants. The reasons for these large differences in susceptibility will be discussed in the next section.

Selective toxicity is very important in both human toxicology and ecotoxicology, but for different reasons. In human toxicology the focus of attention is upon a single species—*Homo sapiens*. Species other than man are used as surrogates in toxicity testing. Thus, when carrying out toxicity tests, there is a great interest in finding test species that are most comparable to humans in their toxic response. Unsurprisingly, primates, the surrogates most closely related to humans, are often regarded as the best models. However, this raises serious ethical problems. There is much opposition to the use of primates in toxicity testing by groups promoting animal welfare, and to a very large extent, toxicity testing is performed upon mammals other than primates, e.g., rats, mice, rabbits, or guinea pigs. Here, there is the difficulty of extrapolating results obtained in this way to humans, of translating toxicity values obtained with rats and mice to estimates of toxicity to humans. There is always the question with such toxicity tests: Which laboratory species best represents humans?

In ecotoxicology the situation is quite different. Tests are performed on a few surrogate species that represent a very large number of wild species—many of which are only very distantly related to the wild species seen to be at the highest risk. Very seldom is the test species the same as the species thought to be at most risk in the natural environment.

Common to both ecotoxicology and human toxicology is the same fundamental question: Which surrogate provides the best model for a particular test chemical? In attempting to resolve this question, we need to delve a little more deeply into the mechanisms that are responsible for toxicity. This will be attempted in Chapter 2.

SUMMARY

Ecotoxicology is here defined as the study of the effects of chemicals upon ecosystems. The ultimate concern is on effects of chemicals at the level of population. Effects upon individuals are important if they are translated into effects at the level of population or above. A central question, then, is: When do effects at the individual level become translated into changes at the population level? Changes at the population level may be in population density or in genetic composition. By contrast, human toxicology (medical toxicology) is particularly concerned with effects upon individuals.

In both cases there are legal requirements for toxicity tests using surrogate organisms that will provide data for the process of risk assessment. In medical toxicology these surrogates are vertebrate animals that are regarded as appropriate models for humans. In ecotoxicology a few surrogate species are used to represent a large number of wild species to which they are seldom very closely related.

The term *pollutant* will be reserved for chemicals for which there is clear evidence of an ability to cause harm at environmentally realistic concentrations. A contaminant is not necessarily a pollutant.

There can be large differences between species, sexes, and age groups regarding their susceptibility to the toxic action of any individual chemical. This raises questions about which species are the most suitable ones to use surrogates in toxicity testing. Pesticides, a group of chemicals of considerable interest to ecotoxicologists, have been classified into groups according to their lethal toxicity to animals (Briggs 1992).

FURTHER READING

Briggs, S.A. 1992. *Basic guide to pesticides: Their characteristics and hazards*. Boca Raton, FL: Taylor & Francis. A useful reference book on the toxicity of pesticides.

Walker, C.H., Sibly, R.M., Hopkin, S.P., and Peakall, D.B. 2012. *Principles of ecotoxicology*. 4th ed. Boca Raton, FL: Taylor & Francis/CRC Press. An introductory text in ecotoxicology based on an MSc course, "Ecotoxicology of Natural Populations."

2 Ecotoxicity

INTRODUCTION

Ecotoxicology is primarily concerned with the effects of chemicals expressed at the levels of population, community, or ecosystem. That said, effects at the level of population and above are usually the consequence of effects upon individuals. Thus, lethal and sublethal effects of pollutants on individuals of a species can lead to population decline if they are severe enough. Such effects can be measured upon live animals and plants in the laboratory. This issue will be discussed further in Chapter 4.

In what follows the term *ecotoxicity* will refer to toxic effects that are relevant in ecotoxicology. Both lethal and sublethal effects of chemicals upon individuals are measured in standard *ecotoxicity tests*, and these procedures will be discussed in more detail later in this section. The results obtained using such tests can be utilized in the process of environmental risk assessment, and this will be described in Chapter 16. At the present time most ecotoxicity tests are performed on living animals or plants, but with advances in biochemical toxicology, there is growing interest in the development of in vitro test systems and model systems such as quantitative structure-activity relationships (QSARs) for estimating ecotoxicity.

One complication in ecotoxicology is that harmful effects upon populations in the natural environment are not necessarily direct; they can be indirect as well. There are complex relationships between different populations in nature, and the effect of a chemical on a population of one species may have knock-on effects upon populations of other species that exist in the same ecosystem, even though they themselves are not directly affected by the chemical. For example, a population decline of one species caused by the toxicity of a pesticide may lead to a related decline in another species that feeds upon it. On agricultural land, for example, the intensive use of herbicides can cause a population decline not only of weeds, but also of insects that feed upon weeds. In turn, this decline of insect populations can cause a related decline of grey partridges (*Perdix perdix*)—because of the dependence of their chicks upon a supply of insect food (see Potts 1986, 2000). These and other examples of indirect effects are very important in ecotoxicology and will be returned to later.

This brings us to a fundamental dilemma in ecotoxicology. Effects at the individual level may or may not lead to consequent effects at higher levels of organization. How can we distinguish between effects that lead to population decline and others that do not? This issue will be discussed further in Chapter 4. Ideally, ecotoxicity tests should give forewarning of potential hazards *at the level of population or above*, under field conditions. Unfortunately, statutory ecotoxicity tests cannot be relied upon to do this.

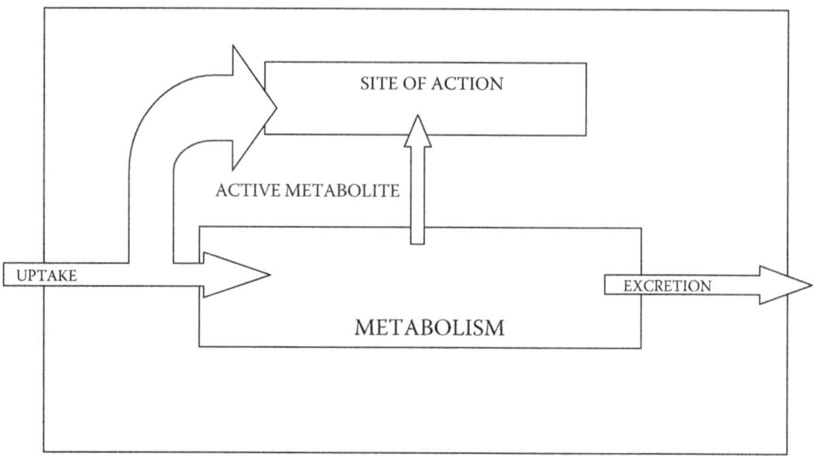

FIGURE 2.1 General model describing the fate of organic pollutants in living organisms.

WHAT DETERMINES ECOTOXICITY?

Why are there large differences in susceptibility to particular chemicals between different groups of organisms? To attempt to answer this question, it is necessary to look a little more closely at the phenomenon of toxicity. Figure 2.1 is a highly simplified representation of the events that occur within a living organism after it is exposed to a chemical that has a significant degree of toxicity. This simple model is based on the processes that occur in animals. However, the picture in plants is essentially similar. Most important, pollutants are fat-soluble (lipophilic) organic compounds, and this model applies particularly to them—although it can, with only minor modification, be applied to other pollutants as well.

When an animal is exposed to a chemical, that chemical may gain entry to the circulatory system and the organs of the body by one or more routes. The most important routes of entry for animals are via the alimentary tract, across the skin or cuticle, through the lungs (terrestrial animals) or the gills (fish). Once inside the animal, it will be carried by the circulatory system to various organs and tissues. Lipophilic pollutants tend to remain in the bodies of terrestrial animals for long periods of time, stored in the fatty tissues (e.g., body fat), unless they are broken down. A very important means of eliminating them, especially in terrestrial animals, is to convert them into water-soluble products that are readily excreted. This important function is carried out by a battery of enzymes that convert them to water-soluble and readily excretable products. Prominent among these are a variety of oxidative enzyme systems that incorporate different forms of cytochrome P450 as their active centers (Box 2.1). These enzyme systems are able to metabolize many fat-soluble molecules that are foreign to the organism in question (xenobiotics). Fish and other aquatic animals are less dependent on this protective system than are terrestrial animals. This is because they can eliminate moderately lipophilic pollutants by diffusion into ambient water (e.g., across the gills)—a route of elimination that is not available to terrestrial animals. These

BOX 2.1 OXIDATIVE ENZYME SYSTEMS OPERATING THROUGH CYTOCHROME P450

The cytochromes P450 are many and varied. They are heme proteins that have the capacity to bind molecular oxygen—rather as hemoglobin does in the blood of vertebrates. Once bound, the oxygen becomes activated by an enzymic process that involves the transfer of electrons. Activated oxygen can then "attack" lipophilic molecules that are bound at a site in the vicinity of the active oxygen. A chemical reaction takes place, transforming the lipophilic molecule into water-soluble products.

The cytochrome P450s are usually located within membranes such as the endoplasmic reticulum of liver cells. Incoming lipophilic pollutants diffuse from the aqueous cytosol into this hydrophobic environment where they can be degraded by these oxidative systems that operate through the cytochromes P450. The products of metabolism (metabolites) are usually water soluble. Consequently, they tend to diffuse out of the lipidic membrane environment into the aqueous cytosol, from whence they are moved out of the cell and are available for excretion. In vertebrates most excretion is via urine or bile.

While this process is usually protective to animals, leading to removal of potentially toxic compounds from the body, it can sometimes misfire. The oxidation of certain carcinogens and organophosphorous insecticides (OPs) causes activation, not detoxication. For examples, see Chapter 10.

***Sources*: Timbrell, J., *Principles of Biochemical Toxicology*, 3rd ed., London: Taylor & Francis, 1999; Walker (2009). Organic Pollutants: An Ecotoxicological Perspective, 2nd ed., Boca Raton: Taylor and Francis.**

detoxifying enzymes tend to be less well represented in fish than they are in terrestrial animals.

Toxic effects are produced when pollutants interact with one or more sites of action within the animal. Many examples of these sites of action will be given in the second part of this text. It is important to emphasize at this early stage that, while many of the toxic effects of pollutants arise because the original molecule interacts with one or more sites of action, in a few, yet critically important, number of cases the molecule that causes the damage is a reactive metabolite. Thus, we can say that the protective metabolism sometimes goes wrong, and the metabolic product causes the damage. This is the case with a number of carcinogens (e.g., benzo(a)pyrene) and some organophosphorous insecticides (e.g., dimethoate and malathion).

PESTICIDES AND OTHER BIOCIDES

Pesticides and other biocides are used with the intention of causing damage to pest species and other organisms that threaten the health or well-being of humans. Because of the potential problems that they present to the environment, pesticides have been subject to more rigorous *ecotoxicity testing* than have most other types of contaminants, as part of the process of *statutory environmental risk assessment.* More ecotoxicity data are usually required about pesticides than for most other industrial chemicals (e.g., pharmaceuticals) that may be released into the environment. On this account, pesticides will be discussed first, before giving attention to other potential contaminants that are not subject to such rigorous testing. With pesticides there are usually clear guidelines as to what uses are permissible—which formulation may be used at which rates and under which circumstances, etc. At least, this is usually the situation in the developed world, where there are normally clear regulations about pesticide use, enforceable by law. In the developing world things are often different. There tend to be far fewer regulations, and those that do exist are not necessarily enforced.

This brings the discussion to a critical issue: Which of the effects that pesticides may have on natural populations are to be seen as acceptable, and which as unacceptable? The control of pests or vectors of disease is clearly both intentional and desirable. On the other hand, the development of resistance in these populations is neither. Resistance appears to be an inevitable outcome of the overuse of pesticides. If the selective pressure of a pesticide is too great over a long period (say ten years of consistent use), the development of resistance in target organisms is to be expected (see Chapter 5). A more controversial question is what effects on nontarget organisms are acceptable. In the first place, harmful effects on populations of beneficial organisms are unacceptable. But there is then the question: Which organisms are to be considered beneficial? Beneficial organisms are taken to include parasites and predators that control pests, also pollinating insects (e.g., bees and hover flies) and soil invertebrates and microorganisms that promote soil fertility. Not included here are the great majority of natural occurring species; these are not usually categorized as beneficial organisms. However, there are also issues about the desirability of biological diversity and the conservation of genes that are potentially beneficial to man, questions that will be dealt with in later chapters.

Sometimes the outcome of statutory risk assessment of a new pesticide, based upon ecotoxicity tests performed in the laboratory, raises questions about possible harmful effects that it may have if used in the field. This concern may arise because of lethal or sublethal effects shown toward test animals. It may then be decided that further testing is necessary before the compound can be approved for marketing by a legislative authority (see Chapter 18). Here, one of the options for the manufacturers of the pesticide may be to carry out field trials to establish whether or not the pesticide does have harmful effects when used in the field (see Somerville and Walker 1990). Unfortunately, field trials are expensive and time-consuming and are seldom carried out in practice. All too often the value of the process of risk assessment is limited because of considerations of cost. This issue will be discussed further in Chapters 4 and 18.

INDUSTRIAL CHEMICALS OTHER THAN PESTICIDES

Returning to environmental chemicals other than pesticides and biocides, statutory requirements for ecotoxicity testing are usually less stringent. Most industrial chemicals are not very toxic to living organisms. That said, pharmaceuticals present a special case because they are, by definition, biologically active. They are used as human or veterinary medicines— e.g., beta blockers, analgesics, antipyretics, diuretics, chemotherapeutic agents, contraceptives, etc. When administered to humans, the doses are usually well below those known to have harmful effects. Typically, only very low concentrations find their way into the environment. However, some problems of pollution have arisen with pharmaceuticals. There has been concern, for example, about a component of the contraceptive pill, ethinylestradiol (EE2), which has been shown to cause endocrine disruption in fish. Also, the anti-inflammatory drug diclofenac has caused serious declines in vulture populations in India when administered to cattle. More generally, questions have been asked about the possible collective effects of pharmaceuticals when they are present in complex mixtures in surface waters—albeit at very low concentrations (see Chapter 19).

These issues aside, pharmaceuticals in general are not usually subject to the rigorous ecotoxicity testing that is required for pesticides and biocides—which are intentionally released into the environment at high enough concentrations to have ecotoxicological effects!

PROTOCOLS FOR ECOTOXICITY TESTING

When determining protocols for the ecotoxicity testing of a new pesticide or biocide, its chemical, physical, and biological properties come into consideration. So too do questions about its release into the environment: where and how it will be released, at what rate, and in what form. These factors can all contribute to the ecological risk.

A fundamental issue in ecotoxicity testing is the range of doses to be given to test species. Doses may be given orally or topically to terrestrial animals. Aquatic species (e.g., fish, amphibians, molluscs) are exposed to a range of concentrations in water. Logically, such tests should cover the range of concentrations likely to be encountered in the living environment, including relatively high levels expected when considering worst-case scenarios. Although this approach is generally accepted, complications can arise when there are legislative requirements for median lethal doses or concentrations (LD_{50} or LC_{50}). Countries differ markedly in their requirements for such data. Sometimes doses need to be given that are far above expected environmental concentrations in order to determine a value for median lethal dose or concentration required by statute. Animals are exposed to concentrations that greatly exceed levels that are likely to occur in the environment—in order to obtain values that are not relevant when estimating environmental risk. Both environmental scientists and campaigners for animal welfare are strongly opposed to this, but legislative requirements for this kind of data still exist. This issue will be discussed further in Chapter 18.

Ecotoxicity tests are carried out upon selected indicator species to yield data that can be used for statutory environmental risk assessment. This process will be discussed in more detail in Chapter 18. In an ideal world these tests would be carried out upon those species deemed to be particularly at risk. However, in the real world, this is seldom possible. A limited number of laboratory species are used as surrogates for the species actually at risk in nature. Typically, only a few species of mammals, birds, fish, or beneficial invertebrates are available for routine ecotoxicity testing. This immediately raises uncertainties about differences in susceptibility between the species seen to be at greatest risk *in the field* and the *surrogates for them* that are used in ecotoxicity testing. A further question is at what stage in its life cycle is an animal or plant most susceptible to the toxic action of a chemical; ideally, ecotoxicity tests should be performed at the most susceptible developmental stage. Questions include: Are young animals more susceptible than older ones? Are the nymphs or larvae of insects more or less susceptible than adults? And, are birds particularly susceptible when the embryo is developing within the eggshell? Questions such as these arise when designing ecotoxicity tests. Unfortunately, because of constraints of cost, availability of skilled labor, etc., much testing is standardized, leaving only limited opportunity for a flexible approach to address these issues.

Another important issue in ecotoxicity testing is which endpoint should be used. Commonly the endpoint used has been death—at least, until recently. Tests have been carried out to determine median lethal dose (e.g., LD_{50}) or median lethal concentration (LC_{50}) for animals maintained in the laboratory. However, as will become apparent later in this text, population declines may be caused by sublethal effects. Effects upon reproduction or behavior can occur at pollutant levels well below median lethal doses or concentrations, and these can cause populations to decline. Examples of this have included the decline of populations of birds of prey due to eggshell thinning caused by p,p′-DDE and the decline of certain molluscs due to loss of fertility caused by tributyl tin. Recently there has been evidence suggesting that populations of bees can decline because of the behavioral effects of neonicotinoid insecticides (Henry et al. 2012; Whitehorn et al. 2012). There is growing interest in designing ecotoxicity tests that will measure relevant sublethal effects of pesticides and other pollutants. Included here are in vitro testing methods using cell cultures and tissue preparations. A lethal toxicity test is rather a blunt instrument, and as knowledge of ecotoxicology advances, it is to be hoped that more sophisticated ecotoxicity tests will emerge. Once again, however, there are limitations of resources and cost when it comes to the development of improved ecotoxicity testing methods.

DETERMINATION OF MEDIAN LETHAL DOSE AND MEDIAN LETHAL CONCENTRATION

The lethal toxicity of a chemical to mammals, birds, and other vertebrates is often expressed as a median lethal dose (LD_{50}). In routine toxicity testing a single oral dose is given to individual animals to obtain a measure of acute oral LD_{50}. Groups of animals are given doses of a test chemical over a range of values that centers on a rough estimate of LD_{50} obtained in preliminary testing. The numbers of individuals that die in each group over a fixed period are recorded, and these values, expressed as percentages, are plotted against the doses, thus generating a *dose-response curve.* The graphical presentation of this is shown in the upper section of Figure 2.2. As can be seen, the graph is curvilinear in character. A statistical transformation of data can now be performed. If the percentages of mortality are converted into probit units, this curvilinear response becomes a straight line, as shown in the lower

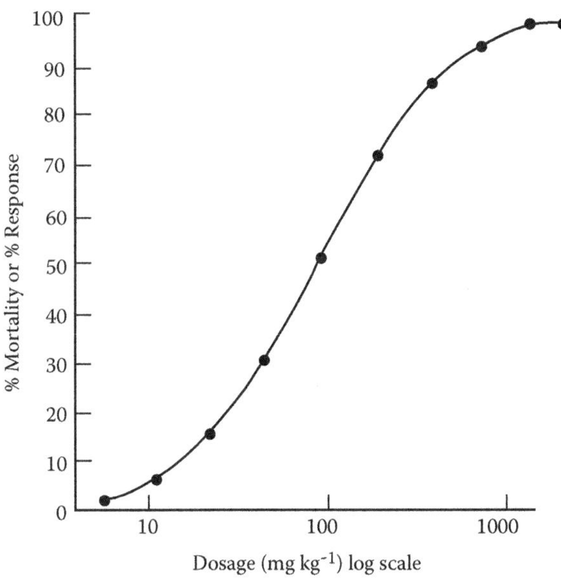

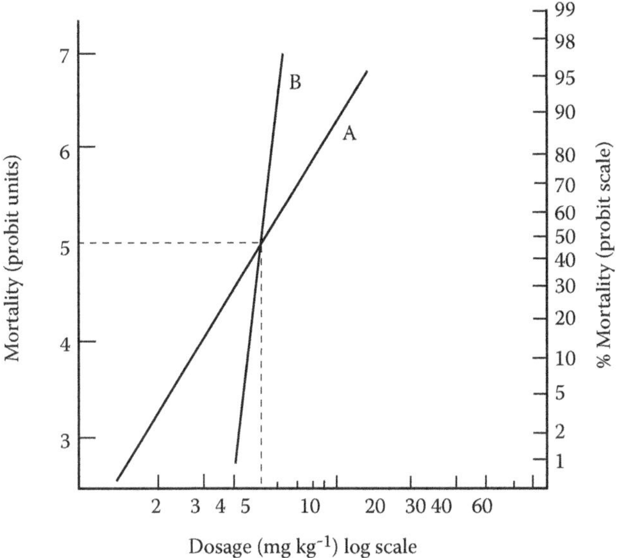

FIGURE 2.2 Determination of LD_{50}. (From Walker, C.H., et al., *Principles of Ecotoxicology*, 4th ed., Boca Raton, FL: Taylor & Francis/CRC, 2012.)

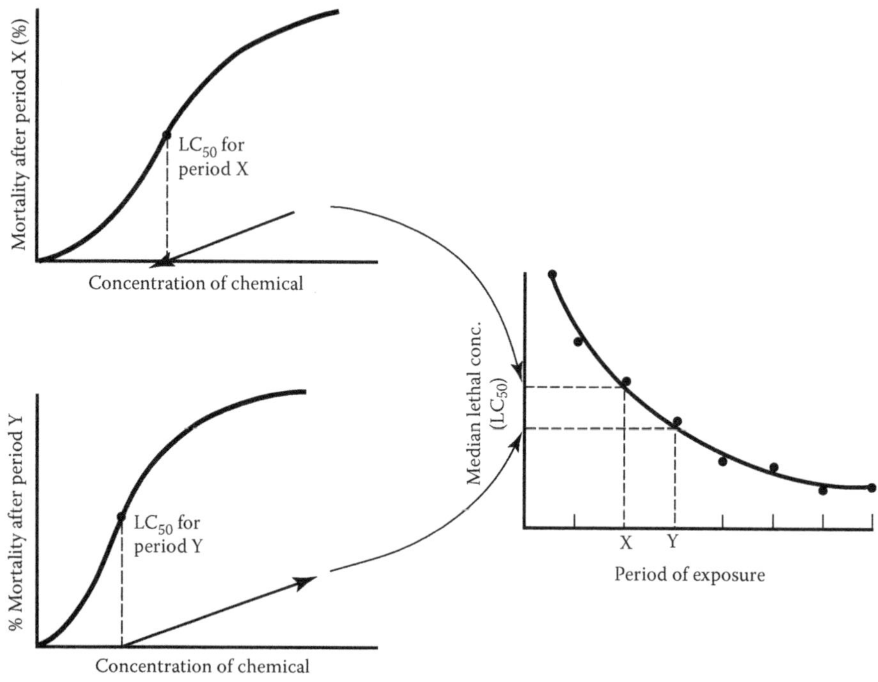

FIGURE 2.3 Determination of LC_{50}. (From Walker, C.H., et al., *Principles of Ecotoxicology*, 4th ed., Boca Raton, FL: Taylor & Francis/CRC, 2012.)

section of Figure 2.2. It is now possible to read from the graph an estimate of LD_{50}, the dose that will lethally poison 50% of the test animals. A statistical transformation of this kind can give a more reliable estimate of values at the extremities of this dose-response curve. More information on this approach can be found in Finney's *Probit Analysis* (1964).

With aquatic organisms such as fish and aquatic invertebrates, lethal toxicity is often expressed as the lethal concentration in the ambient medium, water. Here, there is a different situation from that just described for acute oral toxicity testing which is often based upon the administration of a single dose of a chemical to a test animal. Aquatic organisms are continuously exposed to chemicals present in ambient water, and percentage of mortality will increase over time as the tissue concentrations rise. This is illustrated in Figure 2.3.

After a series of preliminary tests, a range of concentrations of chemicals in water are selected for further testing. Groups of test organisms are then exposed to different concentrations of the test chemical over a period of time, and mortalities are recorded at regular intervals. From these plots estimations of LC_{50} can be made for different periods of exposure to the chemical. In general, the longer the exposure period, the lower the LC_{50}. Important to emphasize here is that the lower the LC_{50} value, the higher the toxicity. With data of this kind, a period of exposure can be chosen for routine testing. Typically, median lethal concentrations for toxicity to fish are measured over a period of ninety-six hours.

BOX 2.2 POTENTIATION AND SYNERGISM

In the natural environment organisms are often exposed to mixtures of environmental chemicals. There is then the question: What is the toxic effect of a mixture of compounds upon an organism? Very often the toxicity of a mixture approximates to the summation of the toxicities of the individual compounds. However, this is not always the case. Taking the example of just two compounds to which an animal is exposed at the same time, if the toxicity significantly exceeds additivity, there is said to be *potentiation*; if, however, the toxicity is significantly less than additive, there is said to be *antagonism*. From a toxicological point of view, significant levels of potentiation in mixtures can be a cause of concern.

A particular case of potentiation is termed *synergism*. Here, a mixture of two compounds is substantially greater than additive—but one of the two compounds in the mixture, the *synergist*, would not be toxic on its own at the dose in question. Frequently a synergist has the effect of inhibiting the metabolism of the other, "toxic" compound.

Synergists of this kind have sometimes been used to enhance the effectiveness of pesticides, e.g., the addition of the synergist piperonyl butoxide to natural pyrethrin insecticide. However, this is not normally practiced today because of unforeseeable risks. A synergist such as piperonyl butoxide may increase the toxicity of chemicals in the environment other than pesticides.

That said, piperonyl butoxide and related inhibitors can be useful in the identification of mechanisms of resistance to insecticides. Frequently, oxidative detoxication by P450-based monooxygenase is a resistance mechanism developed toward pyrethroids and certain other insecticides. Where this is the case resistance will be substantially decreased if this synergist is applied together with the insecticide. Inhibition of the resistance mechanism increases toxicity.

An important point to emphasize is that, up to this point, discussion has been about the toxicity of individual compounds. In nature, however, the situation is more complicated than this, and organisms are commonly exposed to mixtures—and this raises the issue: What is the toxicity of a mixture? This issue is discussed further in Box 2.2.

ECOTOXICITY TESTING THAT USES SUBLETHAL ENDPOINTS

The foregoing two sections have described ecotoxicity testing procedures that utilize lethality as the endpoint. In principle, the same testing protocols can work for sublethal endpoints, many of which have or can be used in ecotoxicity testing. As will become apparent later in this book, sublethal effects, e.g., on reproduction or behavior, may cause populations to decline in the natural environment. Sublethal effects that have been used as endpoints in ecotoxicity testing upon animals have included reductions in reproductive success, the thinning of avian eggshells, changes

BOX 2.3 BIOMARKERS

Biomarkers are here defined as "biological responses to environmental chemicals, at the individual level or below, that demonstrate a departure from normal status." Thus, biochemical, physiological, morphological, and behavioral effects will all be considered as biomarkers. Biological responses at higher levels, e.g., of populations, communities, and ecosystems, are not covered by this definition.

The most valuable biomarkers provide some measure of the progress of a mechanism of toxicity—and will here be termed *mechanistic biomarkers*. Other biomarkers indicate only the presence of a chemical, and do not measure toxic manifestations. These will be termed *biomarkers of exposure*.

Mechanistic biomarkers include inhibition of brain acetyl cholinesterase by organophosphorous insecticides (Chapter 10), eggshell thinning caused by p,p'-DD*E* (Chapter 9), and imposex in bivalve molluscs caused by tributyl tin (Chapters 11 and 15).

Biomarkers of exposure include the production of vitellogenin when assay systems are exposed to estrogens and the induction of certain enzymes (Chapter 15).

Biomarker assays can be used in field studies to give evidence of population declines that are caused by pesticides and other pollutants (see discussion in Chapters 18 and 19).

Sources: **Peakall, D.B., *Animal Biomarkers as Pollution Indicators*, London: Chapman & Hall, 1992; Peakall, D.B., and Shugart, L.R., eds., *Biomarkers: Research and Application in the Field of Environmental Health*, Berlin: Springer Verlag, 1993; Walker, C.H., et al., *Principles of Ecotoxicology*, 4th ed., Boca Raton, FL: Taylor & Francis/ CRC Press. 2012.**

in behavior, neurotoxic effects, and endocrine disturbances. The irony is that these effects can present a greater threat to natural populations than do lethal ones.

Sublethal effects of chemicals upon living organisms can be measured using what have been termed *biomarker assays*. These are described in Box 2.3. Biomarker assays can be used to measure responses of animals and plants using testing procedures that employ sublethal doses of chemicals. From such tests values can be determined for doses corresponding to no effect level, 50% effect, etc.

BIOASSAYS FOR MEASURING TOXICITY

Both cellular systems and genetically manipulated microorganisms have been utilized in the development of bioassays that measure the toxicity of environmental samples. Such bioassays can be used to detect the presence of toxic materials in samples of water, soil, sediment, sewage sludge, air, or material from landfill sites

TABLE 2.1
Some Bioassay Systems

Name of Bioassay	Organism	Principle of Operation	Types of Pollutants Detected	Reference
Microtox	*Vibrio fischerii* A bacterium	Pollutants reduce bioluminescence	Diverse	Persoone et al. 2000
Ames test	Strains of the bacterium *Salmonella typhimurium*	Mutation of microorganism leads to loss of histidine dependence	Various mutagens	Maron and Ames 1983
Calux	Rodent hepatoma cells	Induction of cytochrome P450 1A leads to light emission	Polyhalogenated aromatic hydrocarbons (PHAHs), e.g., dioxin	Garrison et al. 1996
Fish hepatocyte 1	Rainbow trout hepatocytes	Release of the protein vitellogenin	Estrogens	Sumpter and Jopling 1995
Fish hepatocyte 2	Rainbow trout hepatocytes	Induction of cytochrome P450 1A	PHAH, e.g., dioxin	Pesonen et al. 1992

(Lynch and Wiseman 1998). They seldom, if ever, respond only to the toxicity of one particular compound, but they may indicate the presence of a certain type of compound—e.g., an endocrine disruptor or an inducer of a particular enzyme.

They are particularly useful for environmental monitoring—for checking water quality near sewage outfalls or identifying hot spots of pollution in lakes, inland seas, mining areas, and industrial sites. They can measure toxicity caused by mixtures of pollutants rather than individual ones. They can be relatively inexpensive and easy for the nonexpert to use. Thus, with the inevitable constraint of cost, they can identify pollution problems that would remain undetected by more expensive measures such as chemical analysis. Some individual bioassay systems are identified in Table 2.1.

The Microtox assay is nonspecific and responds to a wide range of pollutants. The test organism is the bacterium *Vibrio fischerii*, which emits bioluminescence. A variety of pollutants can have an adverse effect on the bacterium, and this leads to a reduction in bioluminescence. The degree of reduction of bioluminescence provides a measure of toxic effect. It is sensitive to a range of pollutants, but its value is limited because it lacks specificity.

The Ames test has been widely used because it is able to detect mutagens. Apart from its usefulness in environmental monitoring, it has been used to test for the mutagenic properties of industrial chemicals. The organisms used in the Ames test are histidine-dependent strains of the bacterium *Salmonella typhimurium*, i.e., strains of the organism that have a requirement for the presence of the amino acid histidine in the medium in which they grow. When these strains are exposed to certain mutagens, they undergo mutation and the mutant forms do not have histidine dependency; i.e., they will grow in media that do not contain histidine. The number of mutant cells produced after treatment with the chemical can be quantified—and the extent to which mutant forms of the bacterium are generated after treatment with a chemical gives a measure of its mutagenicity.

A valuable feature of the Ames test is that it includes a metabolic activating system containing oxidative enzymes extracted from mammalian liver. Many potent carcinogens, including some of the aromatic hydrocarbons in cigarette smoke, are not mutagens in themselves—but they are converted into highly reactive and unstable metabolites within mammals—and it is these metabolites that damage DNA and initiate cancer. The oxidative enzyme system can form these reactive metabolites within the test system. So, the Ames test not only detects substances that are mutagenic in themselves, but also detects mutagenic metabolites that are generated within the test system.

The Calux system utilizes hepatoma cells of rodents. By a process of genetic engineering a reporter gene for the enzyme *luciferase* has been incorporated into these cells. When certain polyhalogenated aromatic hydrocarbons (PHAHs), such as dioxins or coplanar polychlorinated biphenyls (PCBs), enter them they interact with a binding site called the Ah receptor, which is present in the cytosol (cell fluid). After this the Ah receptor, together with the bound PHAH, passes to the nucleus, where they cause the reporter gene to send a message to the enzyme luciferase, which is also incorporated into this type of cell. When the enzyme receives this message, it emits light. The quantity of light emitted is proportional to the quantity of PHAH that has initiated this sequence of events. In this way the quantity of certain PHAHs in environmental samples can be measured. The use of this system will be discussed further in Chapter 13.

Fish hepatocyte preparations have been used for environmental monitoring in two distinct ways. First, for the detection of estrogens. Estrogens interact with an estrogen receptor—and this leads to the release of the protein vitellogenin. The quantity of vitellogenin released can be measured and is proportional to the quantity of the estrogen present. The second use is for measuring the *induction* of the enzyme cytochrome P450 1A1 by planar molecules such as coplanar PCBs and PHAHs (see mention of Ah receptor above). Induction involves an increase in the quantity of the enzyme, which can be quantified by radio immunoassay. More will be said about induction later in this chapter.

MODEL ECOSYSTEMS

Another approach to ecotoxicity testing is to measure the effects of chemicals upon model ecosystems. These include microcosms and mesocosms. Mesocosms are usually aquatic systems and include ponds, simulated streams, and enclosures within lakes, estuaries, or coastal waters. Both of these types of model systems permit adequate replication when testing the effects of chemicals on the composition of communities, so that detailed statistical analyses can be performed using adequate controls (Caquet et al. 2000; Walker et al. 2012, chap. 14). Thus, they have an advantage over large-scale field trials where replication is difficult, if not impossible. They also have the advantage that the tests can use ecological endpoints as well as toxicological ones. The main disadvantage is that they are model systems that can be difficult to relate to the real world.

ETHICAL ISSUES

Toxicity tests have raised issues of ethical concern. In particular, there has been much criticism of testing methods that cause suffering to vertebrate animals (Walker 2006). The protesters have ranged from responsible organizations like the European Centre for the Validation of Alternative Methods (ECVAM), Interagency Coordinating Committee Validation for Alternative Methods (ICCVAM), and Fund for the Replacement of Animals in Medical Experiments (FRAME) to militant animal rights organizations, members of which have caused damage and intimidated staff at testing laboratories. There is common ground between these responsible organizations and ecotoxicologists who seek more scientific testing methods. As argued earlier, lethal tests are of limited value in ecotoxicology, and there is a case for replacing them with tests that work to critical sublethal endpoints without causing unnecessary suffering. With advancements in certain areas of biochemical toxicology, it should become easier to extrapolate from the results of in vitro tests to expected outcomes in vivo. The development of sophisticated in vitro ecotoxicity tests could satisfy the aims of both ecotoxicologists and those concerned about animal welfare—it could also, in the long term, be more cost-effective than standard ecotoxicity testing methods using live animals.

SUMMARY

The term *ecotoxicity* here refers to toxic effects that are relevant within the field of ecotoxicology. Ecotoxicity tests are performed on free-living species such as earthworms, beetles, freshwater shrimps, molluscs, trout, locusts, and many others. They are also performed upon laboratory species such as rats, mice, feral pigeons, and Japanese quail, which are used as surrogates for free-living species seen to be at risk from environmental chemicals such as pesticides.

Ecotoxicity testing of some kind is often a necessary preliminary to statutory risk assessment of environmental chemicals, i.e., chemicals produced commercially that will be released into the environment. Risk assessment is intended to establish whether it is environmentally safe to do this. The statutory requirements for ecotoxicity testing and accompanying risk assessment vary considerably between countries. Generally speaking, they are more stringent—and more rigorously enforced by developed countries than by developing ones. In general, there are more stringent requirements for the testing of pesticides and biocides than for the general run of industrial chemicals. Occasionally, if risk assessment of a new pesticide or biocide is inconclusive, a field trial may be carried out to establish the environmental safety, or otherwise, of the product.

Ecotoxicity tests performed on animals may work to either the lethal endpoint or sublethal ones. In recent years more attention has been given to sublethal effects than was formerly the case; this has come with the recognition that sublethal effects, e.g., on reproduction or behavior, may cause population declines. There continues to be interest in the development of biomarker assays that can measure the progress of toxic effects.

Bioassays that utilize cellular preparations are useful for biological monitoring and identifying hot spots in the environment that are highly polluted.

FURTHER READING

Calow, P., ed. 1998. *Textbook of ecotoxicology*. Oxford, UK: Blackwell Science. A detailed reference work describing methods of ecotoxicity testing.

Finney, D.J. 1964. *Probit analysis*. 2nd ed. Cambridge: Cambridge University Press. A text describing the technique of probit analysis that is used in the production of dose-response curves.

Persoone, G., Janssen, C., and De Coen, W. 2000. *New microbiotests for routine toxicity screening and biomonitoring*. Dordrecht: Kluwer Academic/Plenum. Describes a wide range of bioassays that have been used to detect pollutants.

Timbrell, J. 1999. *Principles of biochemical toxicology*. 3rd ed. London: Taylor & Francis. A useful textbook on biochemical aspects of toxicology.

3 A History of Chemical Warfare

AN EVOLUTIONARY PERSPECTIVE

When chemical warfare is mentioned, the first thing likely to spring to mind is the production of chemical weapons by the human race for use in warfare; the employment of mustard gas and lewisite during the First World War or, more recently, the use of a nerve gas against Kurds by Saddam Hussein. During the World War II both the Allies and Nazi Germany possessed stockpiles of nerve gases, which, fortunately, were not used at the time. In what follows, the phenomenon of chemical warfare will be considered from a wider perspective.

First, there is good evidence that the use of chemical weapons by humans against other humans goes back much further in time than the wars of the twentieth century. Indigenous peoples of South America, the Indian subcontinent, Africa, and Oceania have traditionally used chemical weapons both to hunt animals and in tribal warfare. Here, the chemicals in question have been naturally occurring ones such as curare and ouabain, which are used as arrow poisons. Curare is present in some tree resins, and ouabain is extracted from certain plants. Toxins produced by animals have also been used for this purpose. Batrachotoxin, for example, is secreted by certain frogs (*Dendrobates* spp.) and has been used by Amerindians as an arrow poison (Table 3.3).

The examples of plant toxins given above are just two among many of plant metabolites that are highly toxic to humans and many other vertebrates. Among them are compounds that have been used to poison humans. An early example involved the Greek philosopher Socrates. He was poisoned with hemlock in 399 BC, an event described by his pupil Plato. A more recent one was the use of ricin by the Bulgarian secret service to carry out a political assassination. Ricin is found in extracts of the castor oil plant (*Ricinus communis*) and is among the most toxic compounds known toward humans. Coniine from hemlock (*Conium maculatum*) and strychnine from the poison nut (*Strychnos nux-vomica*) are other examples of compounds found in plants that are highly toxic to humans. Many other examples of plant toxins are given in Harborne (1993).

Ian Fleming gave a colorful account of a garden of death in the James Bond story *You Only Live Twice*. In the story this is a fictional garden in Japan belonging to Dr. Guntram Shatterhand, one of the reincarnations of Ernst Stavro Blofeld. Contained within it is a rich selection of poisonous plants. The garden is made available to people wishing to commit suicide. Details are given of many plant toxins, including some that are mentioned here.

PLANT TOXINS AS PESTICIDES

The structures of some plant toxins are shown in Figure 3.1. Of these, coniine and strychnine are mentioned above. Some plant toxins have been used not to poison other humans, but in *the war against insect pests and vectors of disease*. Preparations of pyrethrin 1, nicotine, and rotenone, for example, have been marketed as insecticides. The natural product pyrethrum, which contains pyrethrins, is an extract of the dried flowers of *Chrysanthemum* species. As can be seen, the structures of strychnine and rotenone are quite complex, and contain fused ring systems. Strychnine and coniine have heterocyclic rings containing nitrogen.

The highly successful synthetic pyrethroid insecticides, which appeared during the 1960s, were modeled upon the chemical structures of naturally occurring pyrethrins that are extracted from *Chrysanthemum* species (Chapter 12). More recently, the neonicotinoid insecticides, now very widely used, are modeled on the structure of nicotine. This issue will be discussed further later in the present chapter.

A number of examples of natural products that have been used as pesticides, or as models for pesticides, are given in Table 3.1. All of them are poisonous to insects, vertebrate animals, or both.

THE COEVOLUTIONARY "ARMS RACE"

The question arises: Why do plants produce secondary metabolites that are highly toxic to animals, but do not have any obvious role in the normal biochemistry of the plant? It is now widely believed that many of these compounds have appeared as an outcome of a coevolutionary arms race between plants and the animals that feed upon them (see Ehrlich and Raven 1964; Harborne 1993). Animals, in this case, include both insects and vertebrates. Intensive grazing by animals can threaten the survival of plants, and this damage can be limited if the plants produce compounds that are toxic to the animals. In turn, animals have evolved detoxication mechanisms that can break down plant toxins and so limit their potency. One example is provided by certain enzymes that have cytochrome P450 at their active centers. These are described as microsomal monooxygenases (mixed-function oxidases) and can degrade many plant toxins (Scott et al. 1998; Walker 2009). Enzymes of this type have a critical role in the metabolism of organic pollutants, examples of which will be given in Section 2 of the present text.

PLANT TOXINS AS MODELS FOR PESTICIDES

It has been said that there is nothing new under the sun, and it is certainly true that quite a few pesticides are closely related to naturally occurring compounds. Since the early days of pest control, humans have made use of plant toxins that are believed to have evolved during the coevolutionary arms race. The chemicals featured in Table 3.1 fall into this category. Nicotine, which is found in the tobacco plant, has both pharmacological and toxicological properties. This will be discussed further in Chapter 12. Preparations of nicotine were among the earliest insecticides used by man, and the molecular structure was used as a model for the now widely used

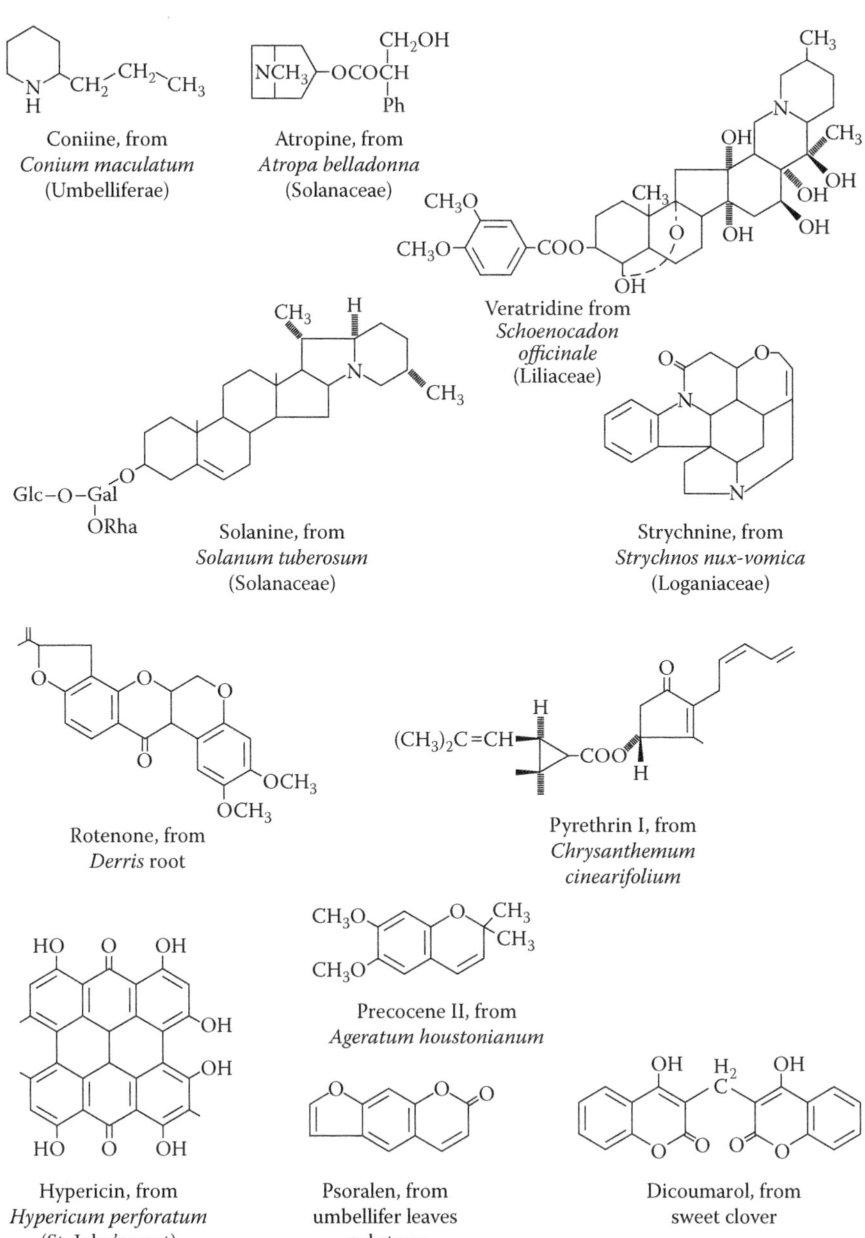

FIGURE 3.1 Some plant toxins.

TABLE 3.1
Some Natural Compounds That Have Been Used as Pesticides

Name	Source	Type of Action	Reference
Nicotine	Tobacco plant *Nicotiana tabacum*	Neurotoxic insecticide	Chapter 12
Pyrethrin	*Chrysanthemum* spp.	Neurotoxic insecticide	Chapter 12
Strychnine	*Strychnos nox-vomica* (plant)	Neurotoxic compound used to control moles	Walker 2009
Rotenone	*Derris ellyptica* (plant)	Mitochondrial poison used as an insecticide	Walker 2009
Coumarol	Sweet clover	Anticoagulant used as a model for rodenticides	Walker 2009
Physostigmine	Calabar bean	Anticholinesterase used as a model for carbamate insecticides	Chapter 10

neonicotinoid insecticides. Natural pyrethrum was once widely used as an insecticide, but was succeeded by the far more effective pyrethroid insecticides, for which it served as a model. Coumarol served as a model for warfarin and related anticoagulantrodenticides. Physostigmine was a model for insecticidal carbamates.

A recurring theme here is the initial discovery of a natural product with pesticidal properties leading to the modification of its structure to produce an economically successful product. In this scenario the original molecule has often turned out to be of only limited efficacy as a pesticide, but subsequent modification of its structure has led to the production of a more effective compound. It can be argued that during the course of the coevolutionary arms race, organisms that are the targets of a chemical weapon have been able to develop defense mechanisms against the toxin. This theory will be discussed further in Chapter 5, which deals with the development of resistance. The important point is this: by judicious molecular design, toxic molecules can be produced that are not subject to defense mechanisms (e.g., enzymic detoxication) that may have evolved during the course of the coevolutionary arms race.

Similar arguments apply to the design of drugs. Many drugs have been designed whose structure is based upon that of natural products that display biological activity of an appropriate kind.

CHEMICAL WEAPONS AND PREDATION

Animals tend to be more mobile than plants—markedly so in terrestrial habitats where plants are usually rooted in the soil. This makes plants vulnerable to grazing by animals—and, as we have seen, there is much evidence that they have evolved chemical defense systems which give some protection against this. By contrast, the mobility of animals provides a mechanism whereby they can avoid being eaten by other animals, a strategy not enjoyed by plants. Despite the advantage of mobility, predation is very common in the animal world, and one might expect to find other

TABLE 3.2
Chemical Weapons of Attack

Group	Description	Mode of Action	Reference
Venomous snakes	Different toxins in venom of different species	Neurotoxicity Cardiotoxicity Blood clotting Hemolysis	Hodgson and Guthrie 1980 Crosby 1998
Scorpions	Toxins in venom	Neurotoxicity	Hodgson and Guthrie 1980
Poisonous spiders	Toxins in venom	Neurotoxicity	Hodgson and Kuhr 1990
Parasitic wasps	Toxins in venom	Cause paralysis	Hodgson and Guthrie 1980

TABLE 3.3
Chemical Weapons of Defense

Organism	Principle	Mode of Action	Reference
Bombardier beetles (*Brachinus* spp.)	Beetle fires a hot noxious cocktail at assailants	Toxic action of quinones generated in an abdominal gland	Agosta 1996
Colombian arrow poison frog (*Phyllobates* spp.)	Frogs secrete the steroid batrachotoxin	Batrachotoxin increases permeability of membranes, thus causing paralysis	Crosby 1998
Cane toad (*Bufo Marinus*)	Poisonous liquid squirted from neck		Agosta 1996
Puffer fish (*Fugu vermicularis*)	Certain tissues contain tetrodotoxin	Tetrodotoxin is a potent neurotoxin	Eldefrawi and Eldefrawi 1990
Fulmar petrel (*Fulmarus glacialis*)	Spits a foul-smelling liquid at assailants	Liquid contains a noxious oil	Cramp et al. 1974

strategies also employed in order to avoid predation. It is now clear that animals as well as plants possess chemical weapons, and these can be for either defense or attack. Predators have weapons of attack; prey have weapons of defense. Some examples of both kinds are shown in Tables 3.2 and 3.3.

CHEMICAL WEAPONS OF ATTACK AND DEFENSE

As noted earlier, indigenous people of Latin America, Africa, and Oceania have long used natural toxins as arrow or dart poisons when they are hunting. These compounds immobilize the prey without necessarily killing it. This is also true of many of the toxins listed in Table 3.2, which are utilized by predatory vertebrates and invertebrates. Venomous snakes, scorpions, and spiders are listed here. With certain parasitic wasps, such as the potter wasp (*Eumenes coarctatus*), small caterpillars paralyzed by the wasp sting are put into the receptacle in which eggs are laid—where they await consumption by the wasp larvae. Immobilization of prey aids predation. A diversity of toxins with contrasting modes of action are represented in this table. More details about them are given in the cited references.

Table 3.3 gives some information about weapons of chemical defense used by animals. The chemicals are diverse—and they are sometimes employed in a spectacular fashion. Bombardier beetles fire a hot noxious mixture of quinones at their attackers. Fulmar petrels are seabirds that spit a foul-smelling oil if they are threatened. Neither of these events necessarily threatens the life of the attacker, but both effectively "see them off."

Much more deadly is the action of tetrodotoxin, which is found in certain tissues of the puffer fish. Puffer fish (*Fugu vermicularis*) are inhabitants of warm seas and are often found on coral reefs. Tetrodotoxin is highly toxic to predatory fish and also to human beings. It is regarded as a delicacy in Japan, where it sometimes appears on the menu of certain restaurants—after careful removal of tissues expected to contain the poison. Despite this precaution there are sometimes cases of fatal poisoning. It has been estimated that there are about 100 fatalities per year due to tetrodotoxin. Tetrodotoxin is an organic cation that can block vital sodium channels in the membranes of nerve and muscle cells—and so cause rapid death.

Chemical weapons like tetrodotoxin are not uncommon in slow-moving herbivorous fish that inhabit enclosed spaces such as reefs. For them, avoidance of predation by rapid and agile movement is not an effective strategy, and chemical weapons can improve their chances of survival. By contrast, chemical weapons of defense are uncommon in fast-swimming fish of the open ocean that rely upon speed and agility to avoid predation.

Frogs and toads are generally slow-moving animals that are vulnerable to predation, and they often use chemical weapons of defense. A range of toxins have been found in this group. Some species have potent poisons in secretions of skin or salivary glands. The cane toad (*Bufo marinus*), an introduced species in Australia, can fire a toxic secretion at perceived assailants—including humans. Indeed, it is regarded as a serious public nuisance in that country.

MICROBIAL TOXINS

Up to this point the discussion has been about chemical weapons that are produced by plants and higher animals. But microorganisms also produce toxins that act against other microorganisms during the struggle for survival, and some of these (e.g., penicillin) have been successfully developed for use as antibiotics in human medicine. This is a wide subject that will only be dealt with briefly here.

Some microbial compounds are highly toxic to human beings and other vertebrates. Many of these are synthesized by fungi (mycotoxins) (Flannigan et al. 1991). The ergot alkaloids constitute one group of such toxins. These are produced by a fungus called *Claviceps purpurea*, which infects grasses and cereals. This fungus produces fruiting bodies called ergots. These contain derivatives of lysergic acid that are poisonous to humans and other vertebrates. These ergot alkaloids are hallucinogenic.

Long before such substances were deliberately consumed by drug addicts, there had been outbreaks of ergot intoxication in human populations exposed to ergot of rye. In the middle ages the phenomenon of ergotism was called St. Anthony's fire.

Ground nuts infected with certain strains of *Aspergillus flavus* or *Aspergillus parasiticus* become contaminated with the hepatic carcinogen aflatoxin B1. This compound, which is synthesized by the fungi, can cause liver cancer in humans. Aflatoxin B1 is also highly toxic to turkeys and has been the cause of heavy mortalities of birds fed contaminated ground nuts.

Another fungal group—species of *Fusarium*—can produce toxins that are highly poisonous to farm animals.

Microorganisms have been a source of natural toxins utilized as pesticides. The bacterium *Bacillus thuringiensis* (B.T.) is an interesting example. This bacterium produces a protein (endotoxin) that contains an insecticidally active element. This natural insecticide can be released by the action of enzymes present in the insect gut. A gene coding for this protein has been incorporated into certain genetically manipulated (GM) crops.

Another example of microbial toxins with insecticidal properties is the avermectins, complex molecules produced by the bacterium *Streptomyces avermitilis*. Eight forms of avermectin are known, and two of these are found in the commercial insecticide abamectin. These and further examples of natural compounds used as pesticides are described by Copping and Duke (2007).

ECOTOXICOLOGY AGAINST THE BACKGROUND OF CHEMICAL WARFARE IN NATURE

A few of the organic chemicals that are used as examples of pollutants later in this text are natural toxins; many more are products of chemical industry that are related in structure to natural toxins. In the latter case, natural toxins have provided models for the synthesis of novel pesticides. As already mentioned, pyrethroid insecticides have been modeled upon natural pyrethrins, and neonicotinoid insecticides upon nicotine. Warfarin and related anticoagulant rodenticides have been modeled upon the natural toxin coumarin. Carbamate insecticides were originally modeled on the natural product physostigmine. Very often the commercial products are not only more effective as pesticides than their natural precursors, but also more likely to have undesirable side effects on nontarget organisms.

The foregoing examples are of pesticides that have been modeled upon natural products derived from plants. Other examples have been of pesticides that utilize microbial toxins. Included here are preparations of *Bacillus thuringiensis* and the avermectins. In all these cases human beings have designed chemical weapons that resemble naturally occurring compounds that have biological activity—and it is now believed that many of these compounds have arisen as a consequence of chemical warfare in the natural environment.

During the course of plant-animal warfare it is believed that defense mechanisms have been developed by animals to protect them against plant toxins; oxidizing enzymes containing cytochrome P450 have been especially important here (see

Lewis 1996). In insects, oxidative enzymes of this type have been found that act against commercial insecticides such as the pyrethroids and the neoniocotinoids. They detoxify these insecticides by converting them into products (metabolites) of low toxicity. Thus, when new insecticides are first used in the field, insects often already possess enzymes that can detoxify them (see Livingstone and Stegeman 1998). It is noteworthy that this happens despite the fact that the enzymes in question cannot have been exposed to these insecticides before. This issue will be discussed further when considering the development of resistance in Chapter 5.

The subject of ecotoxicology is very largely concerned with the harmful effects of man-made chemicals upon the natural environment. However, it is important that this is seen against the background of the history of chemical warfare in the natural environment. Many of the pesticides that we synthesize are closely related to natural toxins, and many of the defense systems that operate against them must have been around long before the advent of *Homo sapiens*. The involvement of the human race in these things is a very recent event on the evolutionary scale.

SUMMARY

There is much evidence that chemical warfare has been taking place in nature over a long period of evolutionary history. The human race has only evolved very recently on this timescale, and humans are relative newcomers to the art of chemical warfare. Indigenous peoples were evidently early users of chemical weapons. Indigenous African, Indian, and Amerindian peoples have evidently used natural toxins for both hunting and warfare since early in human history. Such poisons have often come from plants, sometimes from animals. They have been used as arrow or dart poisons.

Chemical weapons of both attack and defense are relatively common in the animal kingdom. Predators use chemical weapons of attack that can immobilize or kill their prey. Examples include snakes, spiders, and wasps that secrete venoms. Prey use chemical weapons of defense.

Many naturally occurring compounds that are toxic to animals originate from plants. Examples include strychnine, ricin, coniine, and physostigmine. It appears that these compounds have arisen as a consequence of plant-animal warfare during the course of evolutionary history. It is believed that there has been a coevolutionary arms race between plants and animals, during the course of which plants have produced toxins that give them some protection against grazing by animals. Both herbivorous vertebrates and plant-feeding invertebrates would have been involved in grazing—and it is interesting that such animals possess enzymes (e.g., P450-based oxygenases) that can degrade and thereby detoxify plant toxins. Apparently such enzymes are also an outcome of plant-animal warfare.

In recent times humans have developed agents with which to wage chemical warfare of one kind or another. Apart from nerve gases and other chemical warfare agents, there have also been pesticides—insecticides, rodenticides, molluscicides, etc. Pesticides have become increasingly important for the control of agricultural pests and diseases, vectors of disease, and parasites of farm and domestic animals.

Many of the most successful insecticides and rodenticides have been modeled on plant toxins. Mechanisms of toxicity that have evolved during the course of evolutionary history have been successfully exploited by the human race in the development of new pesticides and biocides. In time, resistance to these pesticides has developed, sometimes as a consequence of increased detoxication by the resistant organism. Interestingly, enzymes such as certain cytochrome P450-based oxygenases are sometimes involved in the development of resistance—and these are enzymes that apparently evolved in animals during the course of plant-animal warfare.

Bearing these points in mind, there is merit in seeing the recent problems of pollution of the natural environment by man against the background of the long evolutionary history of chemical warfare on earth.

FURTHER READING

Agosta, W. 1996. *Bombardier beetles and fever trees: A close up look at chemical warfare and signals in animals and plants*. Reading MA: Addison Wesley. A readable account of chemical warfare in nature.

Copping, L.G., and Duke, S.O. 2007. Natural products used commercially as crop protection agents. *Pest Management Science* 63: 524–554. A useful review of biologically active natural products that have been used as pesticides.

Harborne, J.B. 1993. *Introduction to ecological biochemistry*. 4th ed. London: Academic Press. A valuable account of biologically active naturally occurring compounds and the question of plant-animal warfare.

Hodgson, E., and Kuhr, R.J. 1990. *Safer insecticides—Development and use*. New York: Marcel Dekker. Gives examples of natural products that have been used as pesticides or provided models for the development of pesticides.

Lewis, D.F.V. 1996. *Cytochrome P 450s structure, function and mechanism*. London: Taylor & Francis. Includes an account of the evolution of cytochrome P450s. Note especially the evolution of cytochrome P450s family 4.

Walker, C.H. 2009. *Organic pollutants: An ecotoxicological perspective*. 2nd ed. Boca Raton, FL: Taylor & Francis. Chapter 1 deals with chemical warfare and includes a description of the evolution of cytochrome P450s.

4 Toxic Effects at Different Organizational Levels

SEQUENTIAL EFFECTS OF POLLUTANTS

In the first place, a pollutant interacts with an individual organism. If the concentration of the pollutant is high enough, it will have a harmful effect upon that individual. A critical question in ecotoxicology is whether harm to the individual leads to an effect at the population level. Figure 4.1 illustrates the sequential linkage between harmful effects at the individual level and consequent changes at higher levels of biological organization, that is, at the levels of population, community, and ecosystem, respectively (see Ramade 1992).

A harmful effect at the individual level does not necessarily lead to any change at the population level. In ecotoxicology it is important to distinguish between effects upon individuals that are expressed at higher levels and those that are not. Not as easy as it seems! At the next stage there is another question: Does an effect upon one species have knock-on effects upon other species belonging to the same community? Finally, there is the question: What effect does a change at the level of community have on the composition or function of the wider ecosystem? These issues will now be given further consideration.

EFFECTS UPON THE INDIVIDUAL ORGANISM

In the first place, a molecular interaction occurs between an incoming pollutant and cellular structures within the organism. Critically, many toxic effects are due to specific interactions between pollutants and their sites of action within the organisms. Often a site of action is located on a protein. Such a protein may constitute a receptor for a chemical messenger, be part of a pore channel through which there is a regulated flow of ions, or be an enzyme. Sites of action may bind hormones (e.g., estrogen receptors), bind neurotransmitters (e.g., acetylcholine), or be the catalytic centers for enzymes such as acetylcholinsterase, which remediate biochemical reactions. Interactions of this kind will be termed biochemical effects.

These localized biochemical effects often lead to physiological disturbances, e.g., of the brain or nervous system or the blood. Next, the localized disturbances can be disseminated, leading to effects at the level of the whole organism. Effects at the whole organism level include sublethal ones on behavior or reproduction that have the potential to cause population declines, as will be explained later.

In contrast to specific biochemical effects, some pollutants are physical poisons that have rather nonspecific effects upon living organisms if they reach high enough concentrations. These include organic liquids such as ethers, ketones, and alcohols.

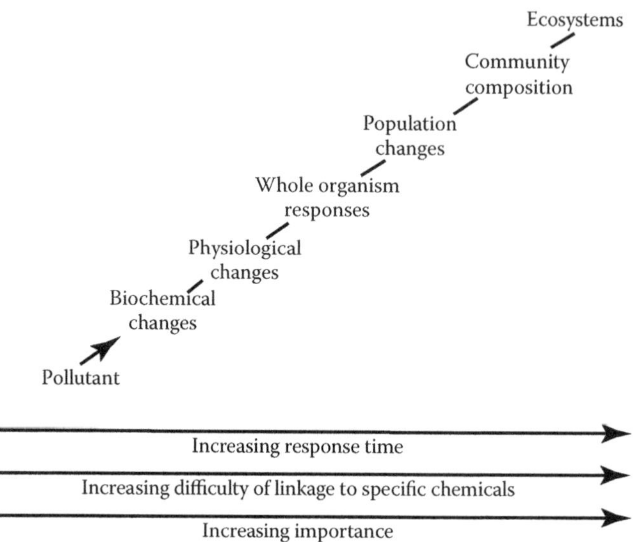

FIGURE 4.1 Schematic representations of linkages between responses at different organizational levels.

If they reach high enough concentrations in biological membranes, they can have physical effects, e.g., changing the membrane permeability.

EFFECTS AT THE POPULATION LEVEL

A population is a group of individuals usually belonging to a single species. Effects at the population level may be characterized in two distinct ways: (1) by changes in population numbers (population dynamics) and (2) by changes in the genetic constitution (population genetics). In the present text, the development of resistance of organisms to pesticides will be given particular attention (see Chapter 5).

Taking population numbers first when studying population dynamics in the field, it is necessary to define the area within which a population is contained. Migration in and out of an area during the course of an experiment can have a profound effect on the population numbers that are recorded. If the effect of a pollutant on a population is to be studied, it is desirable that there be as little migration as possible during the experimental period. It is important not to confuse changes in numbers due to migration with changes caused by chemicals. Commonly, the size of a population varies considerably with time. For example, numbers tend to be high at the end of the breeding season. In northern Europe the population size of resident avian species tends to be high at the end of summer as a consequence of successful breeding, but low at the end of winter due to mortality during the cold weather when food is scarce.

Such fluctuations aside, animal populations tend to increase in size until they approach the *carrying capacity* of the habitat in which they live. The carrying capacity represents a limit to the numbers of any one species that a defined habitat can support. Density-dependent factors such as quantity of food and water, or refuges or

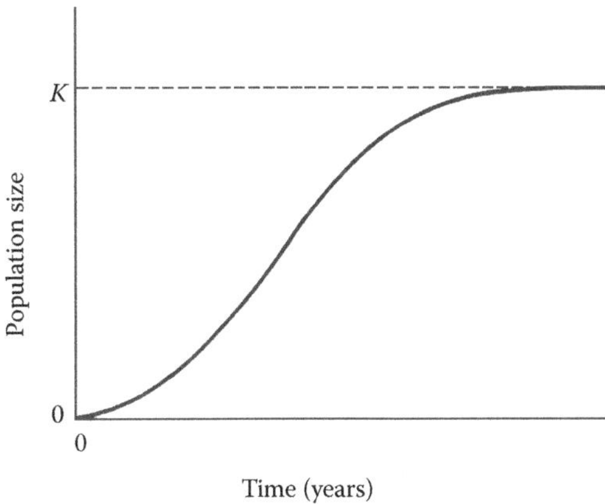

FIGURE 4.2 Increase in population size with time when a species is reintroduced into a habitat.

nesting sites, determine what this limit is. When a population approaches the carrying capacity of a habitat, the population size becomes relatively stable. The rate of loss is more or less balanced by the rate of recruitment. In this situation the loss of a small number of individuals due to the toxic action of a pollutant should not lead to any net decline in population. This is because of the operation of density-dependent factors. The loss of a few individuals due to lethal poisoning can be compensated for by the survival of an equivalent number of individuals that would otherwise have died due to the operation of one or more of these factors; e.g., they would have died of starvation had it not been for the poisoning of some of their competitors. A critical question when animal populations lose members due to lethal poisoning is: How many individuals (or what percentage of the population) need to be poisoned before the population numbers begin to fall?

Another way of looking at this is to consider what happens to an invading species when it moves into a favorable habitat (Figure 4.2). In fact, this represents the situation when a species returns to an area where it once became extinct due to chemical pollution—after the area has become decontaminated (examples of this are given in Chapter 8). Here, the rate of increase in population size is termed the population growth rate. The population growth rate initially increases but later decreases as the population size approaches the carrying capacity of the area. Eventually the population growth rate approaches zero when there is a balance between rate of loss and rate of recruitment. Assuming that there is little migration, the rate of recruitment will largely depend on breeding success. The graph shown is sigmoidal in character.

Population growth rate is a characteristic of central importance in population dynamics. It relates in an interesting way to population size and density-dependent factors. As a population increases in size, so the population growth rate diminishes until a point is reached where the rate is zero. This point represents the carrying capacity of the habitat in which an organism lives. This is the point at which density-dependent factors begin to make their maximal impact. Beyond this point

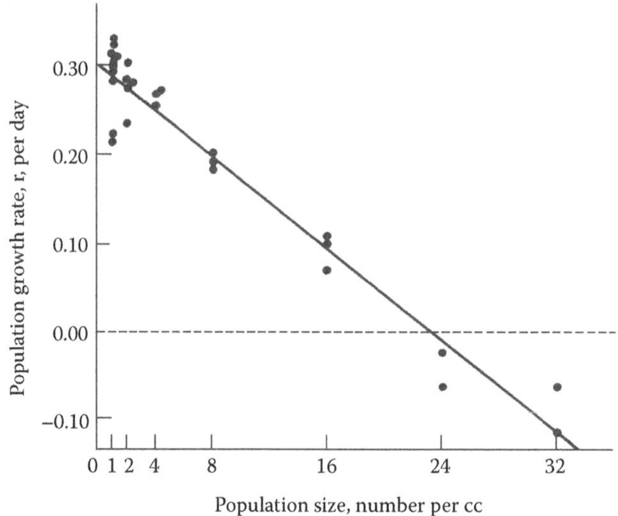

FIGURE 4.3 Relationship between population size and population growth rate.

any further increase in population density is associated with a negative value for population growth rate, i.e., a population decline. See Figure 4.3.

EFFECTS UPON POPULATION GENETICS

Pollutants of all kinds, by the definition given earlier, can cause harm to free- living organisms. They have the potential to exert a selective pressure on free-living organisms. Pesticides, which are designed to control populations of pests, are a very clear example of this. Pollutants in general and pesticides in particular cause stress to free-living organisms and, from an evolutionary point of view, can be likened to other stress factors, such as abnormal regimes of temperature, acidity, and humidity (Van Straalen 2003). Stress, of whatever kind, can be a driving force in the evolutionary process. Thus, the overuse of pesticides can, in the long term, lead to the emergence of resistant strains of pests. This became strikingly apparent with the long-term use of pyrethroids to control pests of cotton, such as the tobacco boll worm (*Heliothis virescens*) in the southern United States (McCaffery 1998). Strains emerged that showed <70,000-fold resistance to a pyrethroid insecticide.

Broadly speaking, the most important resistance mechanisms that have been developed by insects against insecticides are of two kinds: (1) enhanced enzymic detoxication and (2) insensitivity of the site of action. These will be discussed in more detail in Chapter 5.

Resistance to pollutants has also developed in other situations. For example, plants that have been exposed to high concentrations of heavy metals such as copper in mining areas have developed resistance to these metals (Macnair 1987). To widen the perspective, it should also be remembered that pathological microorganisms have developed resistance to antibiotics. The principle is the same. The selective pressure of a harmful chemical causes a change in the genetic constitution of a population.

COMMUNITIES AND ECOSYSTEMS

An *ecosystem* has been defined as a collection of populations that occur in the same place and at the same time that can interact with each other and their physical and chemical surroundings; a *community*, more simply, is a collection of populations that occur in the same place and at the same time (Odum 1971; Calow 1998). Thus, the main difference between them is that *community* is defined simply in biotic terms, while *ecosystem* is defined more widely to encompass interactions with the abiotic environment.

To deal with communities first; these are, typically, rather self-contained areas with definable boundaries, e.g., lakes and soils. Such restricted areas are more suitable for the running of controlled experiments with chemical pollutants than are more extensive ecosystems. Indeed, model systems such as mesocosms have been designed for this purpose, as already mentioned in Chapter 2. Often mesocosms take the form of replicated ponds. Here results obtained from ponds that have been treated with pollutants can be compared with those from control ponds that have not. An example of such a study using a pyrethroid insecticide will be given in Chapter 12. Such replication is very difficult to achieve—often impossible—in the natural environment. Some mesocosms approach natural conditions quite closely. These include enclosures on lakes.

With soils, it is possible to conduct controlled experiments by dividing an area into plots that receive randomized treatments. Test chemicals can be applied at different levels, and some plots can be kept as controls; i.e., they do not receive any chemical at all. Results from properly designed plot experiments can be analyzed statistically to establish whether chemical treatments have had any significant effect upon the composition of the soil community.

With extensive ecosystems, experimental design is much more difficult. The investigator has only limited control over experimental conditions, and satisfactory control areas are often difficult to find. Such was the case with studies on the effects of organochlorine insecticides and other polyhalogenated pollutants on predatory birds in Europe and North America, a subject that will be considered in more detail in Chapters 9 and 13.

EFFECTS UPON STRUCTURE AND FUNCTION OF COMMUNITIES AND ECOSYSTEMS

The effects of pollutants upon communities are of two different kinds. First, there can be effects upon structure (i.e., on composition); here the primary concern is about the species that are present and the numbers of each species. A further aspect is the distribution of sexes or age groups within each species. This subject will be discussed further in the next section.

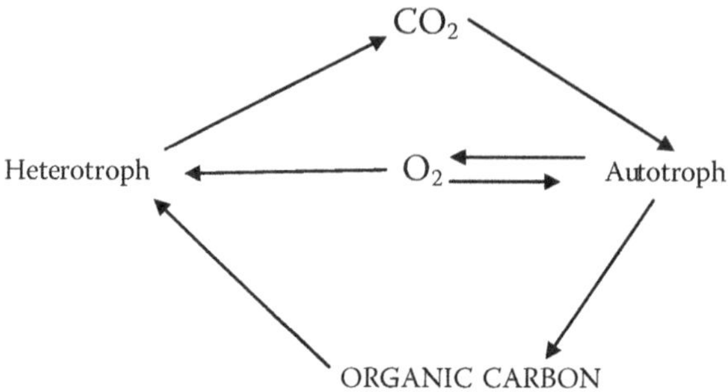

FIGURE 4.4 The carbon cycle.

By contrast, the *function* of a community refers to the operation of processes within it—for example, the operation of the carbon cycle or the nitrogen cycle. Studying these processes can provide measures of the health of a community or of an ecosystem as a whole. Effects of this kind can be measured by monitoring the levels of critical intermediates in natural processes.

At a crude level this principle is illustrated by the effects of raw sewage upon rivers. When such material enters a freshwater system there can be a serious decline in the oxygen levels present in the water. The organic material in the sewage is oxidized by certain bacteria. The bacteria utilize dissolved oxygen to do this—and increase rapidly in numbers. The consequence is a lowering of the oxygen level in the river water. The sewage is said to have a high biological oxygen demand (BOD). This disturbance of a natural process by pollutants can lead to the death of fish due to oxygen starvation.

This effect can be seen in relation to the operation of the carbon cycle (Figure 4.4). Heterotrophs such as aquatic vertebrates and invertebrates and many bacteria oxidize organic compounds in order to release energy, energy that will drive their own metabolic processes. The carbon dioxide so produced becomes available to aquatic heterotrophs such as flowering plants and green algae. These organisms can convert carbon dioxide into sugars by the process of photosynthesis, thus completing the cycle. The raw sewage disturbs the operation of the cycle by facilitating an increase in the numbers of certain types of bacteria that reduce the level of dissolved oxygen in water—an effect that is detrimental to fish. Salmonid fish such as salmon and brown trout are particularly sensitive to low levels of dissolved oxygen. Thus, in this example, a harmful effect of a mixture of pollutants upon an aquatic community (or ecosystem) is indicated by a reduction in the level of dissolved oxygen, a critical intermediate in the operation of the aquatic carbon cycle.

Effects of pollutants on the function of communities or ecosystems can also be observed in soils (see Box 4.1). For example, the herbicide dichlobenil has been shown to increase the rate of production of carbon dioxide in soil (Somerville and Greaves 1987). Effects of biocides upon the process of soil nitrification have also been demonstrated.

BOX 4.1 SOIL ECOSYSTEMS

Soils are complex associations between living organisms and mineral particles. These minerals are bound together by organic polymers to form aggregates that give the soil its structure. These organic polymers are referred to collectively as humic substances (humus) or simply soil organic matter. Humic substances are synthesized in the soil by microorganisms when they decompose organic residues derived from plants and animals. Soils can retain water and plant nutrients such as potassium, phosphates, nitrates, and calcium.

There are many different types of soils, contrasting with one another in their chemical composition, structure, and properties. The latter depends very much on the types of rock from which they are derived and the climate. An important component of soil is the clay mineral fraction. Clay minerals are derived from the weathering of rocks and have small dimensions. Because of this, they have high surface area:volume ratios. In other words, they have large surface areas, and this means that they have a strong capacity to tightly bind certain organic pollutants. Soil humic matter also has strong binding capacity.

In ecotoxicology there is considerable interest in arable soils that have been treated with pesticides, as well as in polluted soils affected by industrial activity. Included here are areas where there are chemical factories, mines, or nuclear power stations. The fate of such pollutants is very dependent upon the type of soil concerned as well as the properties of the chemical. The persistence, movement, and chemical or biochemical transformation of pollutants are all dependent on soil type. Many organic pollutants tend to be more persistent in heavy soils that are high in clay minerals and organic matter than they are in light sandy soils. This is because they tend to be strongly bound to these soil components and are consequently not very mobile. Examples of this include DDT, dieldrin, and other persistent organochlorine insecticides (see Chapter 9 for a more detailed discussion). Many pesticides of higher water solubility tend to be more mobile in soils and are less persistent. Some of these readily percolate through the soil and appear in drainage water, causing pollution problems.

***Source*: Wood, M., Environmental Soil Biology, 2nd ed., Glasgow: Blackie**

MONITORING CHANGES IN THE COMPOSITION OF COMMUNITIES

In ecotoxicology the primary concern is about harmful effects of pollutants that become manifest at the levels of population, community, or ecosystem. Thus, there is particular interest in establishing appropriate monitoring schemes in polluted areas thought to be particularly at risk. Examples include rivers that receive sewage outfalls, effluents from chemical plants, or runoff from mining areas or agricultural land. Soils are also of interest—especially those of agricultural land that are treated with pesticides, and others in the neighborhood of mining operations or chemical plants. In polluted areas there is often some chemical data on the levels of pollutants, which provide valuable background information. However, although such data may set alarm bells ringing, it is often very difficult to establish the biological significance of it. For instance, does a few parts per million (ppm) of a pesticide or heavy metal present any significant threat to invertebrate populations of soil or water?

Biological questions need biological solutions. In the field, bioassays can be used to detect the presence of pollutants that exist at toxic levels (see Chapter 2). These assays are often rather non specific. Many do not give any indication of what sort of toxic action is there. Others, however, have some degree of specificity detecting, for example, neurotoxicity, carcinogenicity, or estrogenic activity. So monitoring polluted areas by chemical analysis or by the use of bioassays can give valuable information on the presence and biological activity of pollutants that are present. But they do not, in themselves, tell us whether the pollutants are having any effects at the level of population or above.

This is where a more ecological approach is necessary. Ideally, a full ecological survey would be carried out, but this is very expensive and time-consuming and can seldom be contemplated. There are, however, biotic indices that can be used to monitor the health of natural populations and communities. These are not prohibitively expensive and can be used to detect changes in the composition of communities that are subject to pollution. A system of this kind developed for fish in the United States is the Index of Biotic Integrity for Aquatic Communities (IBI) described by Karr (1981). Another is the River Invertebrate Prediction and Classification System (RIVPACS), which was developed in the United Kingdom and has been used to assess the quality of UK rivers (Wright 1995). The basic principle of RIVPACS is to establish macroinvertebrate profiles for healthy rivers of different kinds that can then act as controls. These can then be compared with the profiles found in polluted rivers. The comparisons are made between clean and polluted rivers of the same kind, i.e., having similar pH values, oxygen levels, rates of flow temperature, etc. Systems of this kind can, in principle, be used to identify differences between clean and polluted waters with respect to the composition of the communities of aquatic invertebrates or vertebrates that live in them.

FIELD TRIALS

As mentioned in Chapter 1, field trials are sometimes carried out when environmental risk assessment raises questions about the environmental safety of a pesticide. An important issue may be whether a toxic effect found in a laboratory test will be expressed in the field and have harmful effects at the population level in an agricultural ecosystem. The populations under consideration may be protected birds or mammals or beneficial invertebrates. The latter include earthworms, bees, and predatory or parasitic insects that control pests. Such trials are seldom undertaken because they are very expensive to run.

One interesting wide-ranging field experiment on agricultural land was carried out by the Ministry of Agriculture, Fisheries, and Food at Boxworth Farm, Cambridgeshire, UK, during the 1980s (Greig-Smith et al. 1992). The principal aim was to study the side effects of pesticides applied to a cereal crop. A comparison was made of the effects produced using two contrasting regimes: (1) a high-input regime designed to ensure against the occurrence of any pest, weed, or disease problem, and (2) a reduced input system relying on control by the selective use of chemicals and by changes in husbandry with pest control in mind. In other words, chemicals were used tactically, as and when required, and this was supported by modifications of husbandry to facilitate pest control. This study illustrated the difficulty of carrying out such a wide-ranging investigation without adequate controls.

Organophosphorous, pyrethroid, and carbamate insecticides with limited persistence were used for pest control, and a number of short-term population effects were observed—for example, a decline in the population of wood mice (*Apodemus sylvaticus*) that was associated with the use of the molluscicide methiocarb. There were a few examples of long-term declines in the area that were subjected to the high-input regime. In particular two nondispersive ground beetles were severely affected. *Bembidium obtusum* disappeared completely after three years and was still absent at the conclusion of the experiment four years later. *Notiophilus bugattatus* fell drastically in numbers over the first three years and remained at low levels until the end of the experiment. It appears that recolonization of areas treated with insecticide was an important factor here. Nondispersive species like these ground beetles are more vulnerable to repeated use of pesticides of low persistence than are more mobile species. The report of this study gives valuable insights into effects that chemicals can have when used on agricultural land and how these effects may be identified.

SUMMARY

In the natural environment pollutants have effects upon individual organisms that sometimes lead to consequent changes at higher levels of biological organization. When effects occur at the level of the individual, an important question is: To what extent, if at all, are they translated into effects upon populations, communities, and ecosystems?

Effects of pollutants at the population level are of two distinct kinds: (1) changes in numbers (population dynamics) and (2) changes in genetic composition (population genetics). A critical aspect of the latter is the development of resistance to pesticides and other biocides. This can be significant both commercially and medically.

Discrete areas with clear boundaries such as lakes, and certain areas of soil provide examples of communities or ecosystems. Effects at the level of community or ecosystem fall into two categories: (1) effects upon structure, e.g., the relative numbers of different species, or (2) effects upon function, e.g., the operation of natural processes such as the carbon cycle or the nitrogen cycle. Communities and ecosystems are sometimes represented by model systems such as mesocosms, which can be employed to carry out properly designed controlled experiments to determine effects of pollutants. The principal difficulty with this approach is relating results obtained with these model systems to what actually happens in the real world.

FURTHER READING

Begon, M., Mortimer, M., and Thompson, D.J.A. 1996. *Population ecology: A unified study of animals and plants*. 3rd ed. Oxford: Blackwell Scientific. An introduction to population ecology.

Greig-Smith, P., Frampton, G., and Hardy, A.R. 1992. *Pesticides, cereal farming and the environment*. London: HMSO. The official account of the Boxworth experiment.

Odum, E.P. 1971. *Fundamentals of ecology*. Philadelphia: W.B. Saunders Co. A widely read textbook on ecology.

Ramade, F. 1992. *Precis d'ecotoxicology*. Paris: Masson. A wide-ranging text on ecotoxicology that includes a valuable account of mesocosms.

5 The Development of Resistance

INTRODUCTION

Chapter 4 dealt with the effects of pollutants on the size of populations and the composition of communities and ecosystems. We now turn to another issue: the effects of pollutants upon the genetic composition of populations.

POPULATION GENETICS AND EVOLUTIONARY THEORY

Charles Darwin's seminal book *The Origin of Species by Means of Natural Selection* was first published in 1859, with later editions simplifying the title to *The Origin of Species*. In it he expounded a theory of evolution based upon the process of natural selection. Somewhat later the Austrian biologist Gregor Mendel published papers on the patterns of inheritance of certain characteristics in peas between 1865 and 1869, work that was to lay the foundations for the science of genetics. In the long term this work led to the discovery of genes—and genetic selection came to be seen as the basis of natural selection. However, Mendel's work did not come to the notice of the wider scientific community until the twentieth century and was unknown to Darwin. To this day there continues to be a lively debate about the operation of evolutionary theory stimulated by texts such as *The Selfish Gene* by Richard Dawkins (1976).

An aspect of Darwinism that still attracts much interest and public concern is the development of resistance by living organisms to chemicals. By a process of what might be termed unnatural rather than natural selection, certain pathogenic microorganisms have developed resistance to antibiotics and other therapeutic agents to which they have been exposed. Examples of this include the appearance of so-called super bugs that have become resistant to penicillin and other antibiotics to which they have been exposed. An outcome of this has been the difficulty of controlling these pathogens, leading, in some cases, to the closure of hospital wards pending the elimination of resistant organisms. Similarly, the overuse of antimalarial drugs has led to the appearance of resistant strains of the malarial parasite in some parts of the world. Also, strains of the bacterium that causes tuberculosis in humans (*Mycobacterium tuberculosis*) that are resistant to sulfonamide drugs have appeared.

POLLUTANTS AS STRESS FACTORS

Free-living organisms are exposed to stress factors (stressors) such as adverse regimes of temperature, humidity, acidity, and abnormal levels of ions not normally regarded as toxic. These are driving forces in the evolutionary process. Pollutants may also be regarded as

TABLE 5.1
Examples of Insect Resistance to Insecticides

Insecticide	Insect	Use	Resistance Factor	Comment
Cypermethrin, a pyrethroid insecticide	Tobacco bud worm (*Heliothis virescens*)	In cotton crops	< Several hundred-fold	Resistance due to both metabolism and target insensitivity
Various organophosphorous insecticides	Peach-potato aphid (*Myzus persicae*)	In cereal crops		Resistance due to enhanced detoxication by an esterase
Dieldrin	Fruit fly (*Drosophila melanogaster*)	In fruit crops		Target insensitivity
DDT	Housefly (*Musca domestica*)	Fly control in many situations	Kdr < 100-fold Super kdr > 100-fold	Resistance due to both metabolism and target insensitivity

Source: McCaffery, A.R., *Philosophical Transactions of the Royal Society of London B* 353: 1735–1750, 1998; Devonshire, A.L. (1991); Role of esterases in resistance of insects to insecticides. *Biochem. Soc. Trans.* 9: 755–759; Ffrench-Constant, Rochelau, J.C., Streichen, J.G. et al. (1993) A point mutation in a Prosophila GABA receptor that confers insecticide resistance. *Nature* 363:449; Salgado, V.L., in *Pesticide Chemistry and Bioscience—The Food Environment Challenge*, ed. G.T. Brooks and T.R. Roberts, 236–246, Cambridge: Royal Society of Chemistry, 1999.

stress factors (see Korsloot et al. 2004; Van Straalen 2003). One outcome of this can be the emergence of genetic strains well adapted to coping with chemical or other forms of stress. Pollutants can exert a selective pressure upon a population that leads to the emergence of resistant strains. In the ensuing text there will be many examples of this, following the Darwinian principle of evolution as a consequence of natural selection.

Some of the best-studied examples of this have been of insecticides used to control pests of field crops. In the southern United States, for example, heavy use of pyrethroid insecticides on cotton crops led to the development of very high levels of resistance to these compounds in populations of the tobacco bud worm (*Heliothis virescens*) (see Chapter 12). Levels of resistance up to several hundred-fold were found. The dose that controlled susceptible individuals needed to be increased more than 100-fold to control the most highly resistant insects. In this situation it was no longer economic to protect the crop with the insecticide. Some further examples of resistance to insecticides are given in Table 5.1.

MECHANISMS OF RESISTANCE

When pests or pathogenic organisms develop resistance to the chemicals used to control them, ways are sought to get around the problem. In seeking suitable alternatives, a critical question is: By what means was resistance acquired in the first place? That takes us back to a fundamental issue: What factors determine toxicity? (Figure 1.2).

A number of factors have been identified that confer resistance to pesticides, including reduced uptake or storage, changes in metabolism, and insensitivity of site of action. Of these, the most important have been the last two: changes in metabolism and insensitivity of site of action. Some examples are given in Table 5.1

Resistance to dieldrin in strains of the fruit fly has been attributed, largely or entirely, to insensitivity of the site of action, namely, the GABA receptor (Salgado 1999). This is a receptor for the neurotransmitter gamma amino butyric acid, which operates at inhibitory synapses of the nervous system of both vertebrates and invertebrates. Resistant strains of this kind have what is called a mutant gene, a gene that is not found (or is at very low incidence) in susceptible strains. This gene codes for a form of the GABA receptor that is structurally different from the usual form and is insensitive to dieldrin and related cyclodiene insecticides. By contrast, resistance to a range of organophosphorous insecticides (OPs) (see Chapter 10) in peach-potato aphids (*Myzus persicae*) has been attributed to high levels of a detoxifying enzyme in resistant clones (Devonshire 1991). This is an esterase that can degrade and bind the active forms of OPs. Different clones have different levels of this enzyme—and contrasting levels of resistance as a consequence. This phenomenon is particularly interesting from an evolutionary point of view because different resistant clones have different numbers of copies of the same gene and correspondingly differing degrees of resistance to OPs. The numbers of copies of the esterase gene found in resistant clones are multiples of the number of genes in susceptible aphids—2×, 4×, 8×, 16×, etc. The more gene copies in one clone, the greater the resistance to OPs. Thus, the resistance shown by these different clones is not due to the presence of a particular resistance gene in the resistant clones; it is just due to the existence of extra copies of a normal gene that is found in susceptible aphids. Gene duplication has led to the appearance of resistant clones with higher levels of esterase activity than are found in susceptible clones.

Other cases of resistance due to enhanced detoxication have not been related to gene duplication. There has been some evidence for the existence of mutant genes in resistant strains that code for forms of detoxifying enzymes that rapidly degrade insecticides. Also, there has been evidence for overexpression of detoxifying enzymes in some resistant strains. This has been found in strains of *Heliothis virescens* that have developed pyrethroid resistance. For examples, see McCaffery (1998).

In some cases of very high levels of resistance to insecticides more than one mechanism has been implicated. One example is the high levels of resistance to pyrethroids found in strains of *Heliothis virescens* in the cotton fields of the southern United States (Sparks et al. 1993; McCaffery 1998). Levels of resistance up to several hundred-fold have been found in specimens from the field, and levels in excess of 70,000-fold in a field strain that was subjected to selective pressure with pyrethroids in the laboratory. Two contrasting major resistance mechanisms were found to operate in these strains: (1) insensitivity of target site (a sodium channel) and (2) increased oxidative metabolism catalyzed by one or more forms of cytochrome P450. There are parallels here with resistance to DDT in the housefly, where both insensitivity of target site (a sodium channel) and increased metabolic detoxication by DDT-ase accounted for most of the resistance.

CROSS-RESISTANCE

As we have already seen, biologically active natural products have often been used as models for the design of novel pesticides. In general, the biological activity shown by these compounds has depended on the existence of a site of action in pest species with which they—and structurally related synthetic products—can interact. One outcome of this approach to pesticide design has been the emergence of structurally related families of pesticides, members of which share the same mode of action. Examples include pyrethroid, neonicotinoid, and carbamate insecticides and anticoagulant rodenticides. The design of certain new drugs has followed a similar course.

Consider now the development of resistance by a pest species to an insecticide used to control it; this has often been due to the emergence of a genetically distinct strain of the pest that possesses a target site that is insensitive to the insecticide. Now, as we have seen, many insecticides belong to families whose members share the same mode of action. Thus, there is a strong probability that *resistance* due to a mutant gene that encodes for an atypical form of a site of action will extend to different members of the same group of compounds, regardless of whether they have been exposed to the insecticide. In other words, we may expect to find *cross-resistance.*

The appearance of cross-resistance due to the appearance of a mutant gene that encodes for an insensitive form of the target site of an insecticide is a common phenomenon. Cross-resistance has often been found when comparing individual members of the same group of insecticides that share the same mode of action. It can also be found between members of different classes of insecticide if they share the same mode of action. An example of this is the knockdown resistance (kdr) and super knockdown resistance (super kdr) shown by some strains of housefly to DDT (Table 5.1). This type of resistance is due to one or more mutant forms of a sodium channel found in the nervous system. Both DDT and pyrethroid insecticides act upon this site—and strains resistant in this way to DDT also show cross-resistance to pyrethroids (see Chapter 9).

OVERCOMING PROBLEMS OF PESTICIDE RESISTANCE

Prevention is preferable to cure. In the first place, pesticides should be used only when and where they are needed and applied at appropriate dose rates. Unnecessarily high doses will hasten the development of resistance. Sometimes, if the invasion of a crop by an insect pest such as an aphid is noticed early enough, spraying only needs to be carried out over a limited area. At an early stage of infestation, only the margins of a field may be affected by the invading pest.

If there is a serious resistance problem, it may be desirable to use an alternative pesticide that is not affected by the resistance mechanism (or mechanisms) in question. In Chapter 6 there will be discussion about the design of novel pesticides to overcome resistance problems.

As we have seen, some cases of resistance of insects to insecticides have been attributed to enhanced detoxication. Here it may be possible, at least in theory, to counteract resistance by including a synergist in the pesticide formulation that will inhibit detoxication. This can sometimes be achieved with piperonyl butoxide and related compounds that inhibit oxidative detoxification by cytochrome P450s. All very well in theory, but there is a risk that the inhibitors will have unwanted side effects on nontarget organisms—including humans who consume the crop (see Box 2.2).

THE EVOLUTION OF METAL TOLERANCE IN PLANTS

Soils and spoil heaps in the vicinity of mines contain unusually high levels of metals such as copper, tin, lead, zinc, or cadmium. In these locations strains of plants have been found that are said to be tolerant to these high levels of metals. Tolerant strains of plants have been identified that are genetically different from susceptible plants of the same species. In this case, they are comparable to strains of insects that have become resistant to insecticides.

Tolerant strains of plants have been studied in old mining areas in the west of England (Macnair 1987). One site used to be the richest copper mine in the world during the nineteenth century. Here copper tolerance was found in the grass *Agrostis tenuis* within the old mining area but not outside of it. Metal tolerance in plants is discussed in more detail by Moriarty (1999).

RESISTANCE AS AN INDICATOR OF POLLUTION

Earlier in this text examples were given of the development of resistance by insects to insecticides that was characterized by the emergence of resistant strains of pest species. Resistance genes were identified that encoded for insensitive target sites or an increased activity of detoxifying enzymes. A striking case of this was the development of resistance to pyrethroids by strains of the tobacco bud worm in cotton crops in the southern United States earlier in the present chapter. A similar situation apparently exists in the case of plants that develop tolerance to copper and other metals in old mining areas.

In both of these scenarios exposure to a pollutant led to a change in the genetic composition of a population. So, in principle, identifying certain types of genetic change at the population level should be a way of detecting environmental effects of pollutants. It has potential as a means of environmental monitoring. This subject is discussed in more detail in Newman (2010). An example of it is given by Parker and Callaghan (1997). They studied the exposure of blackfly larvae (*Simulium equinum*) in English rivers to low levels of OPs. There was evidence for an elevation of esterase levels in fly populations that had been subject to exposure to OPs over a period of time.

An interesting example of resistance was found in a nonmigratory estuarine fish *Fundulus heteroclitus* inhabiting New Bedford Harbor, Massachusetts, an area polluted by polychlorinated biphenyls (PCBs) since the 1940s (Nacci et al. 2002). These fish were found to have a high level of resistance to a coplanar PCB. As will be explained in Chapter 13, coplanar PCBs, like the related dioxins, are flat molecules that express Ah receptor-mediated toxicity. It would appear that long-term selection by these persistent chlorinated compounds has led to the emergence of a highly resistant strain of a nonmigratory fish.

SUMMARY

Pollutants are an example of stressors—factors that can exert a selective pressure upon natural populations. Following Darwinian principles, selection by them operates in favor of genes that confer resistance to the toxic action of pollutants. Perhaps it would be better to call this unnatural selection, because *pollution*, as defined here, is very largely the consequence of the activities of man.

Pesticides, which are designed to control pests, disease organisms, and other species presenting problems to the human race, are used with the intention of causing ecological damage. When overused, resistance develops to them in the target species. Examples are given here of development of resistance to insecticides by target species. In the course of time, resistance can become so strong that an insecticide loses its effectiveness.

Examples are given of resistance shown by insects to organochlorine, organophosphorous, and pyrethroid insecticides. Two major types of resistance mechanism have been found in resistant strains of insects. First is the appearance of insensitive forms of the site of action. Examples include the enzyme acetylcholinesterase (site of action for organophosphorous insecticides) and sodium channels of the nervous system (sites of action for DDT and pyrethroid insecticides). Second is an increase in enzymic detoxication of insecticides. Oxidative enzymes and esterases are involved here. The appearance of resistant strains of organisms in the natural environment can give evidence of exposure to pollutants.

Strains of plants that are tolerant to metals such as copper have been found in old mining areas.

FURTHER READING

Moriarty, F. 1999. *Ecotoxicology*. 3rd ed. London: Academic Press. A well-written text on ecotoxicology that deals with effects of pollutants upon the genetic constitution of populations.

Nacci, D.E., Coiro, L., Champlin, D., et al. 2002. Predicting the occurrence of genetic adaptation to dioxin-like compounds in populations of the estuarine fish *Fundulus heteroclitus*. *Environmental Toxicology and Chemistry* 21: 1525–1532. An interesting example of a long-term genetic effect attributed to pollutants that interact with the Ah receptor, e.g., coplanar PCBs.

Newman, M.C. 2010. *Fundamentals of ecotoxicology*. 3rd ed. Boca Raton, FL: CRC Press. A well-regarded textbook that deals with the question of effects of pollutants upon populations.

Sparks, T.C., Graves, J.B., and Leonard, B.R. 1993. Insecticide resistance and the tobacco bud worm: Past, present and future. *Reviews in Pesticide Toxicology* 2: 149–183. Contains an account of the development of resistance by *Heliothis virescens* to pyrethroid insecticides.

6 Pesticides and Their Design

INTRODUCTION

During earlier discussion about risk assessment it was explained that pesticides are subject to more rigorous testing procedures than are most other classes of industrial chemicals (Chapter 2). This is hardly surprising since they are, by definition, intended to express environmental toxicity in order to control pests, diseases, and weeds. Many of the pollutants described in this book are pesticides, and the present chapter will be devoted to a brief description of their history before dealing with their properties and design.

EARLY HISTORY OF PESTICIDES

There are early references to chemicals such as sulfur and soda being used as pesticides in classical texts of Greece and Rome, but the first reports of substance date from the nineteenth century. The first records of pyrethrum being used as an insecticide are from this period; also, a few commercial products made their appearance. Paris Green (copper arsenite) was introduced as an insecticide in 1867, and Bordeaux mixture, an insoluble precipitate prepared from copper sulfate solution, was first used as a fungicide (Hassall 1990).

Progress was slow during the early part of the twentieth century. Synthetic mercury fungicides were introduced in Germany in 1913, and the natural product derris powder (also known as rotenone) was first used as an insecticide around 1914.

Between the two world wars a few new pesticides appeared, including dinitro-ortho cresol (DNOC), used as both an herbicide and a fungicide in the 1930s. In 1934 the fungicide thiram was first marketed.

During the World War II events gathered pace. The insecticide DDT, the first of the organochlorine insecticides, was synthesized by Paul Muller of the Swiss firm Geigy in 1939 and was patented in 1942. This came to be widely used by Allied forces to control vectors of disease such as malarial mosquitoes (*Anopheles* spp.). During the war the organophosphorous insecticides (Chapter 10) and the so-called hormone herbicides, e.g., methoxychlorophenoxycetic acid (MCPA) and 2,4-D, were discovered (Chapter 14).

After 1945 there was a revival of the pesticide industry and new products began to appear—so too did some unforeseen problems. In particular, residues of DDT and other persistent organochlorine insecticides in food and biota were seen to present hazards. This was highlighted in the book *Silent Spring* by Rachel Carson, published in 1962. Some persistent residues (e.g., DDE, a metabolite of DDT) and the insecticide dieldrin underwent striking biomagnifications in food webs, and there were side effects upon predators at the top of food chains (Chapter 9). Similar problems were found with persistent organomercury fungicides (Chapter 11), and these too became subject to restrictions and outright bans.

Following these discoveries there was a move toward less persistent pesticides. Persistent organochlorine insecticides were replaced by less persistent organophosphorous and carbamate insecticides (Chapter 10). Organomercury fungicides also were replaced by less persistent alternatives. In time insecticides of limited persistence—at first pyrethroids, then later neonicotinoids—came to dominate the insecticide market. These and more recent developments in the "pesticide story" will be recounted in the second part of the book. In general, there has been a move toward chemicals of greater efficacy and greater safety toward humans and the natural environment. This progress has depended on more sophisticated approaches to pesticide design—a trend that has been made possible by the rapid and remarkable progress in molecular biology and biochemical toxicology over recent decades.

THE IMPORTANCE OF SELECTIVITY

When pesticides are released into the environment, the aim is to control pest species with minimal damage to other organisms. An important factor here is the selectivity of the pesticide. The first concern is with human safety. Spray workers are, so to speak, in the front line. It is clearly undesirable to use pesticides of high toxicity to humans. It is important that workers wear effective protective clothing. Volatile pesticides can be problematic, because they are often readily assimilated via the lungs. If such compounds are used, protective masks need to be worn. But above all, it is desirable that chemicals used are of low toxicity to humans—in the short term and in the long term. There is also the issue of food safety. It is undesirable to have persistent pesticides that give problems of residues in foodstuffs.

Also to be avoided are pesticides that have high toxicity to livestock or domestic animals. This is particularly important in the case of pesticides used for veterinary purposes.

In the field it is vital to have selectivity between target species and beneficial organisms. "What are beneficial organisms?" the reader may ask. In this account the term will include bees and other pollinating insects. Also, natural enemies of pests. The dangers of removing parasites or predators of pest species will become apparent in the later text. It is one way to create a pest problem. A familiar example was the emergence of red spider mite populations as a pest in orchards due to the overuse of insecticides. The trouble was that the insecticides removed capsid bugs, which control red spider mites.

Finally, there is the wider issue of preserving biological diversity. Overuse of pesticides can reduce diversity, and this may have unforeseen consequences. One example is the decline of the grey partridge in Western Europe attributed to highly effective removal of weed species from agricultural land. This will be discussed in Chapter 14. More generally, biological diversity is obviously of interest to naturalists. But it is also, in the wider and the longer term, of interest to the human race more generally. In agriculture the problems of unrestrained monoculture with attendant habitat destruction are coming increasingly to be recognized. One such problem can be the control of pests in the absence of natural enemies. The loss of hedgerows, which provide a habitat for some of them, can be a contributory factor here. Also, the survival of game birds such as the grey partridge on farmland can be threatened if there are too few weeds.

TYPES OF PESTICIDES

Sometimes the term *pesticide* is taken to exclude herbicides. Here it will be employed more widely to include compounds used to control any type of organism that is seen to present a problem or a threat to man, including weeds.

Some examples of types of pesticides are given in Table 6.1. It is not an exhaustive list by any means. It covers types of pesticides that are commonly encountered.

Insecticides of one type or another have long been used on a large scale, particularly in developed countries. They have been, not infrequently, misused in developing countries (Chapter 17). Many problems with insecticide pollution have arisen because they often have considerable toxicity to vertebrates as well as insects. Some insecticides have been problematic because of their high toxicity to humans. In the natural environment there have been many cases of poisoning of mammals, fish, and birds, as will be described later. Rodenticides are far less used, but some of the anticoagulant variety have had a lethal effect on domestic animals—also upon predators such as owls when they feed upon poisoned rodents. There is always a risk that insecticides will have effects on other members of the animal kingdom, because many of them act upon the nervous system, which is similar in vertebrates and insects. Target sites such as acetylcholinesterase, certain sodium channels, and gamma amino butyric acid (GABA) receptors are found in all vertebrate animals.

TABLE 6.1
Examples of Pesticides

Type of Pesticide	Target	Examples
Insecticide	Insect pests; sometimes used to control vectors of disease such as tsetse fly and malarial mosquito	DDT, parathion, malathion, carbaryl, permethrin, imidacloprid
Molluscicide	Molluscs; mainly slugs and snails	Metaldehyde, methiocarb
Acaricide	Acarina (mites); mainly species that damage plants	Cyhexatin
Rodenticide	Rodents; mainly rats and mice	Warfarin, diphenacoum, brodifacoum
Fungicide	Fungi; mainly plant pathogens such as blights, mildews, and molds	Bordeaux mixture, thiram, lime sulfur, EBI fungicides
Herbicide	Weeds	Phenoxy acids, e.g., MCPA, triazines, ureides, herbicidal carbamates
Nematicide	Nematode worms	Some carbamates are used as both nematicides and insecticides, e.g., oxamyl

By contrast, most herbicides show little toxicity toward vertebrates or invertebrates. This is because they act upon target sites, e.g., on the photosynthetic system of the chloroplasts, which do not exist in animals. This is a useful example of selective toxicity, which can be exploited in the design of pesticides. Where a target species has a site of action that does not exist in humans or beneficial organisms, new pesticides can be designed that act upon this site in a pest organism. Such a pesticide should not be very toxic to organisms that lack this site of action.

WAYS OF USING PESTICIDES

Pesticides are employed in a variety of ways. The risks they present to natural populations depend upon the procedures that are followed and the formulations that are used (Box 6.1). In brief, pesticides are applied to crops and fruit trees to protect them against invertebrate (mainly insect) pests and fungal diseases. Application is often to the aerial parts of crops (leaves, stem, flower or fruit). This may be accomplished by using sprays, powders, or dusts. Application may also be to seed or soil. Some pesticides are used to control parasites and diseases of livestock. This may be by oral administration (e.g., helminthicides to control parasitic worms) or as dips to control external parasites (e.g., sheep dips). Sometimes, especially in developing countries, pesticides have been applied aerially from aircraft or helicopters, e.g., when using insecticides to control locusts, malarial mosquitoes, or tsetse flies. In Africa aerial spraying of insecticides has been used to control birds such as quelea.

Aerial spraying is particularly hazardous to wildlife because it is very difficult to control where the pesticide goes. Much depends on atmospheric conditions, especially the strength and direction of wind. There have been cases of pilots being poisoned when flying through their own spray cloud.

Two methods of application deserve special mention. First, the use of insecticidal seed dressings has sometimes caused many deaths of seed-eating birds and mammals. Examples include the organochlorine insecticides dieldrin and heptachlor and the organophosphorous insecticide carbophenothion. Second, the careless disposal of sheep dips containing organochlorine or organophosphorous insecticides into surface waterways has also caused serious pollution problems. These examples will be discussed further in the second section of the book.

OBJECTIVES IN THE DESIGN OF NEW PESTICIDES

To a large extent new pesticides have been developed commercially, and have satisfied certain requirements of the marketplace. As we have seen, there was little of this sort of commercial activity before World War II, but there came rapid and sometimes dramatic innovations afterwards. Much of this became possible because of the striking progress of underlying disciplines such as biochemical toxicology and molecular biology.

In a growing and increasingly competitive market, companies sought to produce pesticides to meet the requirements of their customers. This led to the discovery of new types of pesticides to control pest's diseases and weeds against which existing products were ineffective. There was great interest in the discovery

BOX 6.1 THE FORMULATION OF PESTICIDES

The fate of a pesticide is very dependent on the way in which it is formulated. It has been put this way: "Formulation is a vehicle which allows the active ingredient to be transported to its site of action in the biological system that is to be modified or destroyed" (Hassall 1990). Formulation can affect the toxicity, availability, and persistence of a pesticide, and thus its environmental side effects. The type of formulation chosen for any particular product will depend upon whether the active ingredient is solid or liquid.

LIQUID PESTICIDES

The formulation of a *liquid* pesticide depends upon its solubility. Water-soluble liquid pesticides can usually be marketed as *aqueous solutions*, but pesticides of low water solubility need to be prepared in a different way—usually as an emulsifiable concentrate. In this latter case the pesticide is dissolved in an oily liquid and a surfactant (emulsifier) added. When this preparation is added to water in the spray tank, an emulsion is formed. The pesticide is dispersed throughout the whole volume of water, dissolved in small droplets of the oily liquid.

SOLID PESTICIDES

Solid pesticides can be formulated as dusts or powders and applied as such.

Solid pesticides can also be incorporated into granules, the main body of which is composed of an inert material (filler). Granules have certain advantages over other types of formulation because they can be designed to give sustained long-term release of the pesticide. This feature can have two advantages—long-term control of a pest and only limited availability of the pesticide to spray workers or beneficial organisms. The point is that the bulk of the pesticide is locked up over a long period inside the granule.

A *solid* pesticide may also be formulated as an aqueous solution if it is sufficiently soluble in water. If it has low water solubility, it may be formulated as a dispersible powder or dissolved in an oily liquid and marketed as an emulsifiable concentrate. In all these cases it may be used as a spray.

Please note that in the above account the term *pesticide* refers to the active ingredient and not to the finished product (formulation), which includes other ingredients, such as surfactants, stabilizers, fillers, solvents, etc.

Source: **Hassall, K.A., *The Biochemistry and Uses of Pesticides*, Basingstoke: Hants Macmillan, 1990; Matthews, G.A., *Pesticide Application Methods*, London: Longmans, 1979.**

of compounds that were biologically active—in other words, chemicals that have toxic or pharmacological effects upon living organisms at reasonably low concentrations. Not infrequently these products were based on natural products. The pyrethroid, nicotinoid insecticides, and the anticoagulant rodenticides (e.g., warfarin) are examples of this.

As we have already seen, *selectivity* is an important requirement of any new pesticide. The aim is to maximize toxicity to certain pest species and minimize toxicity to humans, farm and domestic animals, beneficial organisms, and wildlife. Molecular design is of central importance here and will be discussed in more detail later.

In time, some problems arose with existing pesticides that led to innovation and change. The development of *resistance* was one (Chapter 5). An early example of this was knockdown resistance to DDT in houseflies. For a considerable period DDT was used indiscriminately to control houseflies in a variety of situations, including factories, farms, industrial and domestic premises, and places of entertainment. A later example was the development of resistance to pyrethroid insecticides in the tobacco bud worm of cotton-growing areas of the United States (Chapter 5). In response to this problem we have seen both the resurgence of old products and the appearance of new ones that have provided effective alternatives to overcome resistance problems. Overcoming resistance remains a driving force in the design of novel pesticides (see Hodgson and Kuhr 1990).

Another problem has been that of *pesticide safety* in its widest sense. The safety of users of pesticides has been a matter of concern for a long time. There has long been an interest in avoiding the use of compounds that are obviously hazardous. For example, pesticides that have high mammalian lethal toxicity and volatility have been replaced with safer alternatives. Relatively recently other issues have arisen. Some organophosphorous insecticides used for sheep dipping have been associated with a relatively high incidence of neurological disorders in farm workers using them. In response, restrictions have been placed on their use. In general, chemicals shown to have carcinogenic or mutagenic activity have been avoided. In the early screening of candidate new insecticides any compound showing evidence of significant carcinogenic activity is liable to be withdrawn. There has also been concern about the hazards presented to livestock and domestic animals.

The issue of environmental safety is central to the purpose of this book, and is especially tricky, because it is very difficult to establish side effects of pesticides at this level. This is even the case with agricultural ecosystems where it is relatively easy to monitor population numbers and the effects that pesticides have upon living organisms—and is much more difficult with the natural environment if pesticides move into that.

This brings us again to the question of *persistence*. The importance of this became clear when environmental problems associated with the use of persistent organochlorine insecticides and organomercury fungicides were reported during the 1960s (see Carson 1962; Moore 1966; Chapters 9 and 11 in this book). Organochlorine insecticides were detected even in remote corners of the globe, such as Antarctica, having survived transport over many thousands of miles by air masses. Especially worrying was the evidence of their strong biomagnification when they move along food chains, as exemplified in a study upon the Farne Islands ecosystem in the UK where c. 1000-fold higher concentrations of dieldrin and DDE were found in piscivorous birds at the top than in the macrophytes at the bottom of this marine food chain (see Robinson et al. 1967; Chapter 9 in this book). Following these discoveries some persistent pesticides were replaced with less persistent alternatives.

Persistence of chemicals in the environment depends upon the operation of both chemical and biochemical processes. On the one hand, nonbiological processes such as chemical hydrolysis, oxidation, and reduction degrade pollutants at the earth's surface and in water and air. Solar radiation promotes photooxidation. Hydrolysis in surface waters depends upon pH. OPs, for example, tend to hydrolyze more rapidly under alkaline conditions than under neutral ones.

Biological degradation is of central importance in determining the persistence of pollutants in the environment. Fat-soluble (liposoluble) pollutants tend to remain within animals unless they are converted into water-soluble and readily excretable products (Figure 1.2). Most of the early pesticides that caused problems of persistence were heavily halogenated. That means they contained substantial amounts (percentages) of halogen elements in their structures, which gave some protection against enzymic degradation. Usually the halogen was chlorine, and occasionally bromine, iodine, or fluorine. Examples are given in Chapter 9. The more environmentally friendly pesticides that replaced persistent organochlorine compounds were more readily metabolized and more easily excreted.

In designing new compounds that are not unduly persistent, the level of chlorination can be important. These high levels of chlorine in organic molecules tend to make them more difficult to be metabolized by animals, i.e., to be converted biologically into water-soluble and readily excretable products. Dieldrin was one of the most environmentally unfriendly of the organochlorine insecticides, particularly because of its strong persistence. When dieldrin analogues with lower levels of chlorine substitution were synthesized, they were found to be much more rapidly metabolized and excreted than dieldrin itself. This issue will be discussed further in Chapter 9. In general, persistence is a property of pesticides that is very dependent on chemical structure, and one that can be dealt with by changing molecular design.

The persistence of a pesticide depends not only on the properties of the chemical itself, but also upon the way in which it is formulated. Granular formulations of pesticides, for example, can be prepared in such a way as to regulate the rate of release into the environment (see Box 6.1). In this way, a pesticide that is readily degraded in the environment either chemically or biologically can be slowly released over a long period to control a pest. While molecules of the pesticide are locked up inside the granule, they are protected from, for example, biological degradation or the action of solar radiation. Once released, they are subject to the normal processes of degradation and dispersion.

Metabolism can be a critical determinant of selective toxicity. Returning to the model given in Figure 1.2, rapid metabolism to form products (metabolites) of low toxicity minimizes the quantity of a toxic pesticide reaching its site of action. Differing metabolic capabilities between species, strains, sexes, and age groups can be exploited to achieve selectivity. This has sometimes been a feature of pesticide design. For example, the OP malathion contains a carboxy ester bond, which has been termed a selectophore (see Chapter 10). The purpose of introducing this bond into the molecule was to reduce the toxicity to vertebrates, including man. Vertebrates contain certain esterases (enzymes that catalyze the degradation of esters) that attack this bond and facilitate the rapid degradation of malathion to nontoxic products, thereby reducing toxicity. Consequently, malathion is one of the OPs showing relatively low toxicity to humans, and is sometimes used in medicine to control parasitic invertebrates.

MASS SCREENING

Designing new pesticides that will satisfy the demands of the market is not a simple matter. Finding novel compounds that possess the appropriate chemical and biological properties to control an insect pest, a fungal disease, or a weed under field conditions is difficult enough in itself. But on top of that consideration has to be given to operator and consumer safety, avoidance of undesired side effects on beneficial organisms, the agroecosystem, and the natural environment more generally—quite a tall order.

Chemical companies that introduce new pesticides have long relied on a screening process whereby a large number of candidate compounds are subjected to a series of tests to establish their suitability or otherwise for further development as products on the world market. Large numbers of chemicals are tested for their toxicity toward different target organisms—insect pests, plant diseases, and agricultural weeds, for example. From the initial screening a few compounds are selected for further testing.

In earlier times this approach was not very efficient. The initial criteria for selecting the compounds to be tested were not very scientific. Some estimates suggest that several thousand different compounds needed to be screened in order to find one compound that could be successfully developed and marketed. At that time there was a conservative attitude toward new scientific approaches (Hodgson and Kuhr 1990). However, with the progress of science in this area attitudes changed. At Rothamsted Experimental Station, for example, Elliott and colleagues synthesized 100 or so novel pyrethroids, out of which seven were successfully marketed. These included the insecticides permethrin, cypermethrin, and decamethrin. The insecticide structures were based upon those of naturally occurring pyrethrins that have insecticidal activity. The new synthetic compounds followed the geometry of the natural models (Elliott 1977; Leahy 1985; Hassall 1990). Achievements such as these led to more attention being given to the rational design of pesticides and less dependence on random screening.

MODELING OF NEW PESTICIDES BASED UPON THE STRUCTURES OF SITES OF ACTION

Some fundamental advances in biochemical toxicology and pharmacology came during the later part of the twentieth century. Critical here was the elucidation of the 3D structures of the sites of action of pesticides and drugs and the presentation of these on the screen using computer technology. Using these models, it became

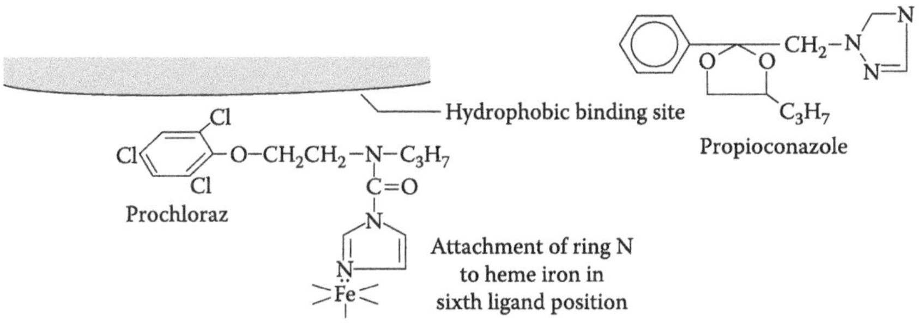

FIGURE 6.1 Interaction of an EBI fungicide with cytochrome P450.

possible to design new biologically active molecules able to interact with these sites. Molecules could be designed to inhibit enzymes or to block sites of binding for chemical messengers such as acetylcholine or GABA. An example of this is given in Figure 6.1.

The design of a number of ergosterol biosynthesis inhibitor (EBI) fungicides was guided by models of the active center of a form of the hemeprotein cytochrome P450, which is illustrated in Figure 6.1. This form of cytochrome P450 catalyzes the oxidation of lanosterol, which is a precursor in the biochemical synthesis of the important molecule ergosterol. Ergosterol is required by many fungi to form stable cell membranes. If synthesis of it is blocked, cell membranes can become unstable. The EBIs block synthesis by strongly binding to the active center of the enzyme. Critically, a nitrogen atom that is part of a heterocyclic ring attaches to the iron atom located at the center of the action. This is where molecular oxygen binds and is activated. A hydrophobic (i.e., nonpolar) part of the fungicide attaches to a neighboring hydrophobic binding site of this cytochrome P450, where the sterol precursor would normally be attached. So there is a strong binding of the fungicide to two sites—preventing the attachment of the molecular oxygen or the sterol precursor of ergosterol to either of them. In this way the biosynthesis of ergosterol is inhibited by the fungicide. The EBI fungicide shown in Figure 6.1 is prochloraz.

This and some other examples of types of pesticides for which the structure of the site of action is known are given in Table 6.2. The important point is that these structures, when presented three-dimensionally, can act as templates for the design of novel pesticides. If such compounds bind tightly to these sites, they can have toxic effects because they disrupt living processes. In the example given, the EBIs and OPs act as enzyme inhibitors, and so interfere with natural biochemical processes. DDT and pyrethroids attach themselves to a site on a sodium channel that spans the axonal membrane of nerve cells (neurons). The effect of this is to disrupt the electrical signals that are sent along the nerve. The way this happens will be explained in Chapter 9. Neonicotinoid insecticides also act upon the nervous system, but in a different way from DDT or pyrethroids. Nerves end at junctions that are called synapses. On the other side of synapses are cells of nerves, muscles, or glands. Messages are carried from nerve endings across synapses by neurotransmitters. These neurotransmitters combine with receptors on the postsynaptic membranes of cells of

TABLE 6.2
Models of Sites of Action

Type of Pesticide	Examples	Site of Action	Reference
EBI fungicides	Prochloraz, ketoconazole, flutriafol	A fungal cytochrome P450	Lyr 1987
Neonicotinoid insecticides	Imidacloprid, thiacloprid	Acetylcholine receptors on cholinergic synapses	Jeschke and Nauen 2008
Organophosphorous insecticides	Diazinon, dimethoate, malathion	Acetylcholinesterase	Sussman et al. 1991
Pyrethroids and DDT	DDT, permethrin, cypermethrin	Sodium channel of nerve axon	Hodgson and Kuhr 1990

a nerve, muscle, or gland, and by doing so pass on a message originating from the nerve cell. Acetylcholine is such a transmitter. On cholinergic synapses acetylcholine released from a nerve ending crosses a synapse and interacts with a cholinergic receptor of a muscle or gland cell on the other side. Neonicotinoids act as antagonists of nicotinic acetylcholine receptors, and so disrupt the transmission of nervous impulses across this synapse.

Models of sites of action for pesticide design come in different forms. There are solid models constructed, typically, from plastic spheres that can be linked together. These are available in various kits. There are also computer models where 3D simulations are displayed on the screen.

Whatever sort of model is used, constructs of novel pesticides can be made in the same modular system, and then tested against the site of action to establish the goodness of fit, and how complementary their geometry really is. New compounds that fit the model can then be synthesized and tested for their toxicity toward target organisms.

Modeling the design of new pesticides is potentially useful when seeking to overcome a resistance problem. As we have seen, resistance of insects to insecticides is frequently due, at least in part, to the possession by a resistant strain of an insensitive form of the site of action. Examples of this include mutant forms of sodium channels of the nervous system (as in kdr and super kdr resistance to DDT) or of acetylcholine esterase (Salgado 1999; Devonshire et al. 1998).

Some of the resistant forms of acetylcholinesterase differ from normal forms of the enzyme by a change in only one of the many constitutive amino acids. However, molecular models show that even a change in a single amino acid can create an obstruction to the action of an OP insecticide. An abnormal amino acid can prevent an OP from reaching the active center of the enzyme. More about this phenomenon will be said in Chapter 10. The critical point here is that modeling can, in principle, lead to the discovery and development of new pesticides that will overcome resistance. In the present example resistance might be overcome by the design of a new OP that, on account of its geometry, would not be subject to steric hindrance by the abnormal amino acid of a resistant form of acetylcholinesterase.

SUMMARY

In the present account, the term *pesticide* refers to herbicides as well as insecticides, fungicides, rodenticides, molluscicides, etc. Pesticides are specifically intended to have harmful effects against target organisms, and they are consequently subject to more rigorous testing than most industrial chemicals to establish the risks that they pose when released into the environment. Pesticides are formulated in a number of different ways. Emulsifiable concentrates and wettable powders can be dispersed in water as a preliminary to spraying. Pesticides are also formulated as dusts and granules. The manner of formulation affects the persistence and distribution of a pesticide in the environment, and thus the environmental risk that it poses.

Some of the earliest pesticides were natural products, such as nicotine, pyrethrum, and rotenone used as insecticides, which came into use during the early part of the twentieth century. However, it was not until after World War II that the pesticide industry began to develop on a large scale, with the marketing of many new synthetic compounds. By the early 1960s the alarm bells began to ring about the possible side effects of persistent organochlorine insecticides and methylmercury fungicides on the natural environment. Evidence grew that some of these chemicals were undergoing marked biomagnification with movement along food chains, reaching especially high concentrations in predators of the highest trophic levels.

Of particular concern were persistent organochlorine insecticides (OCs) such as DDT, aldrin, dieldrin, and heptachlor, and certain of their metabolites (e.g., DDE, a persistent metabolite of DDT). Aldrin and dieldrin were implicated in declines of the sparrowhawk and peregrine in the United Kingdom and other states of Western Europe. DDE was found to be responsible for eggshell thinning and population declines of predatory birds, e.g., the bald eagle, the double-crested cormorant, and the brown pelican in North America. Following these findings, less persistent insecticides, e.g., certain organophosphorous, carbamate, and pyrethroid insecticides, were brought in to replace persistent OCs.

Another problem has been the development of resistance by pest species when pesticides are overused.

With rapid scientific and industrial progress, new, more environmentally friendly pesticides have been introduced in recent times. The rational design of new pesticides has made great progress—and there is less dependence on the mass screening of large numbers of different molecules in the quest for novel compounds. Rational design offers a way of overcoming problems of resistance by the discovery of new pesticides that are not vulnerable to old resistance mechanisms.

FURTHER READING

Elliott, M. 1977. Synthetic pyrethroids. *American Chemical Society Symposium Series* 42: 1–28. A firsthand account of the synthetic pyrethroids.

Hassall, K.A. 1990. *The biochemistry and uses of pesticides*. Basingstoke: Hants Macmillan. A detailed and informative textbook on pesticides.

Jeschke, P., and Nauen, R. 2008. Neonicotinoids from zero to hero in insecticide chemistry. *Pest Management Science* 64: 1084–1098. Describes the properties of the neonicotinoids, a group of insecticides that have come into prominence in the twentieth century.

Leahy, J.P., ed. 1985. *The pyrethroid insecticides*. London: Taylor & Francis. A valuable reference work on the synthetic pyrethroids.

Lyr, H., ed. 1987. *Modern selective fungicides*. England: Harlow Longman Scientific and Technical. A useful source of information on the EBI fungicides.

Moriarty, F., ed. 1975. *Organochlorine insecticides: Persistent organic pollutants*. London: Academic Press. A collection of specialized chapters on the ecotoxicology of the organochlorine insecticides.

7 Natural Pollutants and Natural Cycles

INTRODUCTION

The expression *natural pollutants* may sound like a contradiction in terms. Most substances regarded as pollutants are man-made. However, in Chapter 1 the definition given of *pollutants* included natural chemicals when they occur at abnormally high levels. Very often this happens because of the activities of man—for example, due to mining for minerals, drilling for oil, burning coal, smelting ores to release metals, using internal combustion engines, or operating nuclear power plants. There are also situations where natural pollution occurs outside of any human activity. Cataclysmic events such as the eruption of volcanoes, tsunamis, spontaneous forest fires, or the landing of meteorites can all cause pollution in this sense and have serious and widespread ecological effects. Some examples of this have been seen within historical time on volcanic islands such as those of the Galapagos, Hawaii, and Tristan da Cunha.

If we leave aside, for the moment, the question of chemical warfare in nature, there are quite a few naturally occurring chemicals regarded as pollutants, and many of these are inorganic. Included here are metals such as copper, lead, cadmium, and nickel. Toxic gases such as sulfur dioxide (SO_2), nitrogen dioxide (NO_2), and hydrogen sulfide (H_2S) also fall into this category. There are also some naturally occurring pollutants that are organic. The combustion of trees and other vegetation leads to the formation of a wide range of organic compounds, including polycyclic aromatic hydrocarbons such as benzo(a)pyrene and benzanthracene. Benzo(a)pyrene is a potent carcinogen and is one of the chemicals in cigarette smoke that is implicated in the development of lung cancer by humans. Some organometallic compounds also occur naturally. Both methylmercury and methylarsenic compounds are found in marine organisms, including some seafood. They can be synthesized in the environment by certain aquatic microorganisms from inorganic mercury or arsenic. More will be said about organometallics in Chapter 11.

Organic pollutants that are both man-made and naturally occurring can raise problems of interpretation for ecotoxicologists. When they are detected in environmental samples by analysts, questions arise about their origins. Does a residue found in an aquatic organism originate from a natural or a human source? Not infrequently the answer is from both. This leads to the next question: How much is man-made and how much is natural? A critical question if the aim is to control pollution.

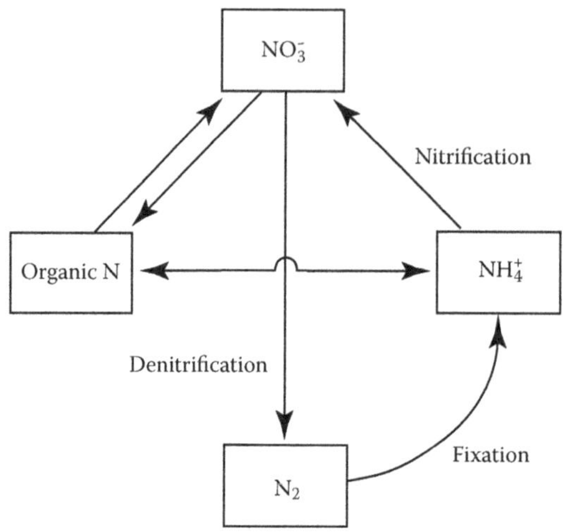

FIGURE 7.1 The operation of the nitrogen cycle.

EUTROPHICATION

Some of the above issues are exemplified by the phenomenon of eutrophication (Nosengo 2005). Without the intervention of man, nitrate ions are generated by the operation of the nitrogen cycle, shown in Figure 7.1.

When organic residues decompose in soils, ammonium ions (NH_4^+) are released, which are eventually converted into nitrate ions (NO_3^-) by microbial action. Nitrate ions readily elute through soils, eventually appearing in drainage waters and so finding their way into neighboring lakes and water courses. The use of commercially produced nitrogen fertilizers has the same effect, and can greatly increase the quantity of nitrate ions reaching nearby surface waters. High levels of nitrate cause rapid increases in the size of aquatic algal populations. Consequently, algal blooms appear that cause environmental problems. When algae die, bacteria break down the dead tissues—leading to oxygen deficiency in the ambient water. This can lead to the death of fish species, e.g., salmonids, which require relatively high levels of oxygen in water (see Chapter 4 and Figure 4.4). In fact, dead algae can have a similar effect to raw sewage in surface waters (Chapter 8).

This scenario may be enacted without the intervention of man. However, the chances of it happening are greatly increased where there is overuse of nitrogen fertilizers. In particular, the nitrate ion passes freely through the soil profile and into surface waters.

ACID RAIN

From a chemical point of view water is neutral at a pH value of 7.0. Above this value it is alkaline; below this value it is acidic. The lower the pH value, the higher the acidity (see Box 7.1). Normal rainwater at equilibrium with atmospheric carbon

BOX 7.1 PH AND ACID RAIN

pH is a logarithmic representation of the concentration of hydrogen ions in an aqueous solution. If we consider absolutely pure water, the following reversible equilibrium exists:

$$H^+ + OH^- \rightleftarrows H_2O$$

H^+ is the symbol for the hydrogen ion; OH^- is the symbol for the hydroxyl ion. When equilibrium exists, both ions are at the same concentration, which is 10^{-7} molar (10^{-7} M). This is an inconvenient number to work with, and the degree of acidity is usually expressed as its *negative* logarithm to the base 10, which is 7. In other words, at neutrality pH = 7.

H^+ ions confer acidity; OH^- ions confer alkalinity. pH values below 7 represent hydrogen ion concentrations above 10^{-7} molar. Solutions of pH below 7 are said to be acidic. pH values above 7 represent hydrogen ion concentrations below 10^{-7} molar. Solutions of pH above 7 are said to be basic basic or alkaline. Such use of a negative value for a logarithm can be confusing.

In practice, rainwater is acidic, and it normally has a pH of around 5. The acidity is mainly due to the carbon dioxide dissolved in it. Smaller contributions come from sulfur dioxide and nitrogen oxides that are normally present in air. When there is pollution by these gases from industrial sources, the pH of rainwater falls. Acid rain can have pH values of 4.2 and below.

Buffer solutions have reserve acidity and alkalinity; this means that they are less affected by acids or bases than is pure or relatively clean water. Certain naturally occurring organic compounds, which are weak acids or bases, act as buffers. Relatively clean natural waters often lack buffering capacity—and so are particularly vulnerable to the action of acid rain. Such is the case with certain lakes in southern Norway and Sweden.

dioxide (CO_2) has a pH of about 5.6, and is therefore acidic, in a purely chemical sense. Carbon dioxide dissolves in water to form carbonic acid, and the acid dissociates to release hydrogen ions (H^+). From the point of view of ecotoxicology, rain is described as being acidic when it has a pH below 5.6, and this is the terminology that will be adopted here.

Rain becomes acidic in this latter sense when it dissolves gases present in the atmosphere—especially sulfur dioxide (SO_2) and nitrogen oxides such as nitrogen dioxide (NO_2). When these gases dissolve, strong acids are formed, e.g., sulfurous acid, sulfuric acid, and nitric acid. They release hydrogen ions and the pH of the rain falls. A contribution to acidity in rain can also come from ammonium ions (NH_4^+) that are formed from ammonia gas (NH_3), itself an air pollutant.

Sulfur dioxide is implicated in the production of acid rain (Howells 1995). Indeed, this gas used to be a serious pollutant in many industrial regions of the world. Sulfur occurs in coal and is at particularly high levels in limonite—the brown coal found, for example, in the north of Bohemia and the south of Poland. In communist times there was serious deforestation of these areas due to the toxicity of this gas to conifers and other types of trees. Large quantities of sulfur dioxide were formed with the burning of the brown coal. Since 1989 this problem has largely disappeared, partly because of the establishment of filters in the chimneys of furnaces and power plants.

Returning to the question of acid rain, a major concern in the northern hemisphere has been its role in the acidification of lakes and other surface waters. In this zone there are many lakes devoid of fish in which the water is poorly buffered (see Box 7.1) and the pH is very low. Examples of this have been reported from Canada, the United States, Norway, and Sweden.

RADIATION AND RADIOISOTOPES

Living organisms are exposed to different types of naturally occurring radiation. Included here are ultraviolet violet (UV) and infrared (IR) radiation originating from the sun. Closer to home, so to speak, there is exposure to radiation emitted by naturally occurring radioactive minerals found in some geological deposits that emit alpha, beta, or gamma rays.

Large doses of radiation can be damaging to living organisms. In humans there are health risks associated with exposure to high doses of solar radiation, e.g., the skin cancer melanoma. Certain forms of bone cancer and acute leukemia can be caused by radiation from atomic explosions, as became apparent after Hiroshima and Nagasaki were destroyed by atomic bombs in 1945 at the end of the World War II. Less dramatic than this, humans and other living organisms are regularly exposed to low levels of radiation that arise from radioactive minerals occurring in local rock formations, and questions have been asked about the health hazards involved.

Minerals are radioactive because they contain unstable radioisotopes that emit radioactivity as they decay. Two types of radioactive emission, alpha and gamma, are featured in Box 7.2. Another type is beta radiation.

The igneous rock granite is a source of many different minerals and also of radioactivity. In some granitic areas of Great Britain, inhabitants are exposed to significantly higher levels of radiation than the British population more generally. Radium raises particular pollution problems because, as it undergoes radioactive decay, the gas radon is released—and radon is itself radioactive. It can enter buildings, thus exposing the inhabitants to dangerous levels of radiation.

Apart from exposure to naturally occurring levels of radiation, there are more serious radiation risks for most forms of life as a consequence of human activity. The principal examples of this are related to the development of nuclear weapons and the operation (or more accurately, mismanagement) of nuclear power stations. Reference has already been made to the catastrophic damage caused by the two atomic bombs dropped on Japan in 1945. There is continuing concern about the possibility of nuclear war now that a number of countries have atomic weapons.

There have been serious pollution problems with radiochemicals in areas that were once testing grounds for nuclear weapons. At one such testing ground at Semipalatinsk in the Kuzbass and Altai regions of the former Soviet Union there was significant radiopollution resulting from 470 nuclear explosions carried out between

BOX 7.2 RADIOACTIVE EMANATION

1. A radioactive isomer of potassium (^{40}K) emits gamma radiation:

$$^{40}K \rightarrow {}^{40}Ar + \text{gamma radiation}$$

2. A radioactive isomer of uranium (^{238}U) emits alpha radiation:

$$^{238}U \rightarrow {}^{234}Th + \text{alpha radiation}$$

3. A radioactive isomer of radium (^{226}Ra) emits alpha radiation:

$$^{226}Ra \rightarrow {}^{222}Rn + \text{alpha radiation}$$

The products of these reactions are isomers of argon (Ar), thorium (Th), and radon (Rn).

Alpha radiation is a stream of alpha particles that are, chemically, nuclei of the element helium having an atomic weight of 4 (2 protons + 2 neutrons). Thus, the atomic weights of the isomers of thorium and radon formed during radioactive decay are each 4 units fewer than those of the isomers of uranium and radium from which they are derived.

1949 and 1962, 87 of which had been aerial explosions. Possible health effects upon the human population of areas in the vicinity of the testing ground were subsequently investigated (see Ilyinskikh et al. 1999). There were indications of genetic damage, including a high incidence of micronucleated cells and Epstein-Barr virus in inhabitants of the area. Unfortunately, this region was also affected by pollution from neighboring metallurgical and coal industries, and it was difficult to distinguish between possible effects of industrial pollutants and those of radiochemicals.

The hazards associated with nuclear power stations when they are not properly managed became all too clear with nuclear disasters at Chernobyl and Chelyabinsk in the former Soviet Union. Chelyabinsk received little attention from the rest of the world when it happened, but Chernobyl gained widespread publicity. At Chernobyl in 1986 a nuclear reactor caught fire and released some 50% of its contents into the atmosphere (Edwards 1994). People living in the area around the site were evacuated, and fallout of nucleotides like ^{137}caesium arising from the disaster was recorded. Some 70% of this fallout occurred in Byelorussia, but large areas of Western Europe were also affected. Sweden was badly affected, and subsequently a study of pollution by the radioisotope ^{137}Cs was conducted there (Ahman and Ahman 1994). Reindeer showed a marked seasonal fluctuation in their body content of this radioisotope, high in winter but low in summer. This was because of a marked seasonal variation in their food. In the winter they consumed large quantities of lichens, which had a high level of this radioisotope, but in summer they ate mainly other things, e.g., grasses, which had a much lower content of it.

After people were evacuated from the area around Chernobyl, an ecological study was conducted. A bizarre conclusion of this was that some animal populations increased after the departure of humans—despite high levels of radiation. The destructive capacity of the human race should not be underestimated.

Chelyabinsk is a region of Russia close to Kazakhstan. This region, with a capital of the same name, used to be a production center for weapons-grade plutonium. It now hosts a large nuclear fuel processing center (Osborn 2011). It experienced at least three serious nuclear accidents during the Soviet era, leading to contamination of a large area. These included a large nuclear explosion in 1957 and were covered up at the time, only becoming more widely known after the end of communism. There was widespread dumping of nuclear wastes into local lakes and rivers, making it a highly radioactive area. There have been reports of high levels of diseases related to radiation among inhabitants of the area.

The foregoing examples of serious nuclear accidents were fairly and squarely the outcome of human incompetence and mismanagement. More recently nature played a large part in a serious case of radiopollution. When the coast of Honshu, Japan, was struck by a big tsunami in 2011, damage caused to the reactor of a nuclear power station located on the coast led to a leak of radioisotopes into the sea. Questions were asked about the wisdom of locating nuclear reactors on coasts with a history of tsunamis.

OIL POLLUTION

Pollution by crude oil has received much publicity. The wreckages of oil tankers such as the *Torrey Canyon* off the coast of Cornwall, UK, in 1967, the *Amoco Cadiz* off the Breton coast of France in 1978, and the *Exxon Valdez* off the coast of Alaska in 1989 caused extensive pollution and were widely reported in the world media. There has also been serious marine pollution as a consequence of leakage and spills at oil terminals, such as the recent very serious incident at an oil installation belonging to British Petroleum (BP), which caused widespread pollution of the Gulf of Mexico. Oil pollution of land has also been problematic because of leakages from storage tanks and road accidents involving oil tankers. There are huge quantities of natural crude oil, and accidents occurring when handling it are liable to be large ones.

Marine pollution by oil and the effects that it can have on some marine vertebrates are familiar to most people. Oiled seabirds demonstrate clearly the damage that crude oil can cause. Apart from these overt effects, which are relatively short term, there are other, more insidious consequences of oil pollution that can be longer lasting. Crude oil contains small quantities of what are termed polycyclic aromatic hydrocarbons (PAHs)—a group of chemicals that includes carcinogens such as benzo(a)pyrene. These compounds are persistent in sediments and in marine invertebrates and other organisms low in the marine food chain. It is unclear, however, whether they have had important carcinogenic or mutagenic effects on marine organisms.

One study looked at the physiological state of edible mussels (*Mytilus edulis*) near an oil terminal at Sullom Voe in the Shetland Islands (Livingstone et al. 1988). The mussels were sampled along a pollution gradient in the neighborhood of the oil terminal, and certain biological indicators suggested that adverse effects increased with progression up the gradient. This study was illustrative of the difficulties of conducting studies in the field. Without further work, it leaves open the question of whether the association between the gradient of aromatic hydrocarbons and the state of health of mussels was a causal one—or whether some other environmental factor determined the physiological state of the mussels.

Relatively recently there have been reports from the west of England of incidents involving hundreds of seabirds washed up dead, following exposure to polyisobutene (PIB). PIB is a synthetic oil used to thicken engine oil. Large oceangoing vessels are permitted to discharge quantities of this at sea. At the time of writing there is an ongoing debate about the question of banning discharges of this chemical at sea.

GREENHOUSE GASES AND GLOBAL WARMING

At first it may seem bizarre to regard CO_2 as a pollutant. After all, it is an essential constituent of the carbon cycle. During photosynthesis plants utilize it as a carbon source in the synthesis of sugars, a vital natural process upon which survival of most living organisms depends. Following the definition given earlier, CO_2, like any other environmental chemical, may be regarded as a pollutant if it exists at an abnormally high level and causes harm to natural populations. Carbon dioxide is the most important of what have been termed greenhouse gases—gases present in the atmosphere that increase the retention of solar energy, and thereby increase the temperature. Some estimates of the contribution of different gases to this greenhouse effect are given in Table 7.1.

Of these gases water vapor makes the largest contribution to the retention of heat by the atmosphere. Carbon dioxide makes a substantial and rising contribution. Of lesser importance are methane and ozone. Ozone (O_3), however, has another, more important influence on the global environment. It is generated from oxygen in the stratosphere and forms a protective layer (shield) above the earth. This ozone layer restricts the entry of potentially harmful solar ultraviolet radiation to the atmosphere. In the latter part of the twentieth century a hole appeared in this layer in the region of the South Pole. A little more will be said about this later in this chapter.

Over the last four to five decades the CO_2 content of the atmosphere has increased by approximately 10%—and the average earth surface temperature has risen by approximately 0.8°C over a similar period (UNEP Environmental Programme). The

TABLE 7.1
Some Greenhouse Gases

Gas	% Contribution to Global Warming
Water vapor	36–70
Carbon dioxide	9–26
Methane	9
Ozone	3–7

connection between these two trends has been the subject of lively debate. Is this really a causal relationship? There have been big long-term changes in the global climate before. The recent ice age is an example. Also, the output of energy by the sun is critical in determining global climate—and there are known to be cycles of energy emission from the sun. However, expert analysis suggests that the recent increase in global temperature does not relate to the operation of solar cycles, but it does correlate well with an increase in atmospheric CO_2.

This rise in global temperature may seem very small, but if it continues, there will be catastrophic consequences in the longer term. The melting of polar ice is an important example. This is already happening, and if it continues, large-scale flooding of low-lying land is to be expected. A sustained rise in global temperature would be expected to cause changes in global weather patterns. Indeed, there are already signs that this may be happening.

THE OZONE LAYER

Ozone is a gas formed when molecular oxygen (O_2) is exposed to certain forms of radiation, including ultraviolet light (see Crosby 1998; Andrews et al. 1996). A thin layer of ozone exists above the earth in the stratosphere at an altitude of 30–40 km (18–24 miles). It is readily formed in the region above the equator, from whence it diffuses out in the direction of the polar regions of the earth. Although relatively thin, this layer has been likened to a protective shield, because it absorbs much of the potentially damaging UV radiation coming from the sun. Indeed, solar radiation would destroy much of the life on the planet were it not for the intervention of the ozone layer.

There was consternation when scientists of the British Antarctic Survey working in the Antarctic reported that a hole had appeared in the ozone layer above the region of the south pole in 1984. Since then, further work has produced evidence that pollutants called chlorofluorocarbons (CFCs) exist in the ozone layer, and that these have been responsible for destruction of ozone. These volatile chemicals have been used as refrigerants and aerosol propellants, and when released at or near the earth's surface, they can make their way by diffusion into the stratosphere. Since these discoveries, bans have been placed on the manufacture of these chemicals, and there has been some recovery of the ozone layer.

Thus, we have something of a dilemma in classifying ozone. Closer to the earth it answers to the description of a pollutant. It is a greenhouse gas to a small extent; also, when present in polluted urban environments it can contribute to the formation of photochemical smog (see Andrews et al. 1996). However, in the stratosphere it forms a protective shield for life on earth. Considerations such as these have encouraged the development of global concepts of pollution, as will be discussed in the next section.

GLOBAL PROCESSES AND THE GAIA THEORY

In the foregoing sections there has been frequent mention of natural cycles such as the carbon cycle and the nitrogen cycle. These have been described as biogeochemical cycles, a term which recognizes that they fall within the boundaries of both biology and geochemistry. The intermediates in these processes may be derived from both the chemical decomposition of minerals and the metabolism of living organisms. CO_2, for example, may be formed by the action of acids of rain upon limestone rock—or by the biochemical degradation of dead plants or animals by microorganisms. Nitrate ions in surface waters may arise from the degradation of minerals or the operation of the nitrogen cycle.

Examples have been given where the operation of these cycles can be disturbed by pollutants. Certain pesticides, for example, can affect the functioning of the nitrogen cycle in soils. Also, we have seen that the operation of the carbon cycle in freshwater can be disrupted by both sewage and excess nitrate. The second example refers to the phenomenon of eutrophication where the nitrate may originate from both normal decomposition of organic residues in soil and the use of inorganic fertilizers.

Returning to a matter of definition as used here, the term *pollutant* is not restricted to compounds synthesized by man, and it may also refer to naturally occurring compounds when they occur at abnormally high concentrations that cause ecological damage. Included here may be volcanic gases, e.g., SO_2, and metals like copper, nickel, or cadmium—or even CO_2 in the atmosphere if the level is too high. Pollution can be the result of the dysfunction of natural processes.

Much of the earlier discussion focused upon aquatic and terrestrial ecosystems—on lakes, rivers, and soils. But when considering acid rain or the effects of CFCs on the ozone layer, the debate moves on to a global scale. The atmosphere constitutes a bridge between ecosystems at the surface of the earth; it acts as a connection between landmasses. If we take further the example of CFCs, these are volatile industrial chemicals designed to act as refrigerants or aerosol propellants. When they came into widespread use, significant concentrations of them found their way into the ozone layer of the stratosphere. Once there, they broke down a proportion of the ozone and created a hole in the layer. This allowed more solar UV light to come through and reach the earth's surface. If this process had continued, there could have been disastrous effects upon life on earth—upon both natural ecosystems and the human race. This is an example of pollution on the global scale—and there is a need for models on the global scale to address such problems.

The air masses of the lower atmosphere (up to c. 35 km above the earth's surface) undergo regular circulation (see Lutgens and Tarbuck 1992). Once pollutants reach a level of c. 4 km above the earth's surface, they enter a circulatory system and may be transported for thousands of kilometers before they are deposited on the earth's surface by falling rain or snow. A striking example of this was observed in 1883 after the huge volcanic eruption at Krakatoa, an island in what is now Indonesia. For months after the eruption volcanic dust contaminated the atmosphere and was widely circulated above the earth's surface. Unusually vivid sunsets were noticed in many parts of the world due to refraction of sunlight high in the atmosphere.

Circulation of air masses has also been important in the movement of gaseous pollutants over considerable distances. For example, some of the SO_2 causing acid rain in Norway and Sweden was evidently exported from chimneys in Britain. Similarly, a pesticide not used in Canada was apparently exported aerially to Lake Superior from the American Midwest. Pollutants do not respect international boundaries. There are some significant political issues in ecotoxicology.

Global pollution problems call for a holistic, global approach. One example has been the development of the Gaia hypothesis (see Lovelock 1982, 1988). The term *Gaia* is taken from the name of a Greek goddess personifying the earth, an "earth mother." James Lovelock, Lynn Margulis, and some others have suggested that the earth acts as a single living entity rather than a randomly driven geochemical system. It is seen as a self-regulating whole within which biotic and abiotic factors interact to produce a kind of homeostasis. Biochemical as well as physical and chemical processes make vital contributions to the cycles discussed earlier, cycles that are biogeochemical ones. Chemical pollution is seen as being the consequence of natural processes as well as the activities of man. A holistic global model of this kind incorporates the phenomenon of global warming, the action of CFCs upon the ozone layer and consequences thereof, and the transport of pollutants (natural or man-made), including acid rain. The concept of Gaia has been widely publicized, generated much learned debate, and raised issues of philosophical interest.

SUMMARY

As defined here, the term *pollutant* includes not only the synthetic products of man, but also naturally occurring chemicals, when they occur at abnormally high levels. In many cases the levels of such chemicals are elevated as a consequence of human activity, e.g., due to mining, burning of fuels, or the operation of industrial processes such as smelting ores containing copper, tin, or lead and nuclear power plants. However, in some cases the same chemicals may appear at abnormally high levels as a consequence of natural processes. Extreme events such as volcanic eruptions, earthquakes, tsunamis, or the landing of meteorites can all lead to the elevation of environmental levels of gases such as sulfur dioxide or metals such as copper.

Many natural pollutants are inorganic in character, including metals such as copper, lead, and mercury and gases such as sulfur dioxide and nitrogen oxides. A few are organic. Included here are certain products of combustion. Extreme events such as volcanic action can initiate forest and bush fires, and these lead to pollution that is the consequence of incomplete combustion of organic compounds. PAHs such as the carcinogen benzo(a)pyrene can be formed in this way. Methylmercury and methylarsenic are also generated naturally due to the activity of microorganisms. Thus, some pollutants originate from both human activity and natural processes.

Natural processes, including the carbon cycle and the nitrogen cycle, can be disturbed by naturally occurring compounds. Thus, excess nitrate in surface water can lead to eutrophication. On a larger scale, too much carbon dioxide in the atmosphere can cause global warming, which can have harmful consequences for the living environment. Also, due to human activity the thinning of the ozone layer caused by CFCs has raised concerns about an increase in UV radiation reaching the earth's surface, and resultant harmful effects upon living organisms. Considerations of this kind led to the proposal of global models such as Gaia, which regard the earth and its atmosphere as a single, self-regulating entity. Global models of this kind can give insight into global pollution problems.

FURTHER READING

Andrews, J.E., Brimblecombe, P., Jickells, T.D., et al. 1996. *An introduction to environmental chemistry*. Oxford: Blackwell Sciences. A textbook that deals with global warming, the thinning of the ozone layer, and biogeochemical cycles in a clear and concise way.

Edwards, T. 1994. Chernobyl. *National Geographic* 186: 100–115. An account of the nuclear accident at Chernobyl.

Howells, G. 1995. *Acid rain and acid waters*. 2nd ed. Hemel Hempstead, UK: Ellis Horwood. An authoritative text on acid rain.

Ilyinskikh, E.N., Ilyinskikh, N.N., and Ilyinskikh, I.N. 1999. In *Biomarkers: A pragmatic basis for remediation of severe pollution in Eastern Europe*, ed. D.B. Peakall, C.H. Walker, and P. Migula. NATO Science Series Environmental Security, vol. 54. Dordrecht: Kluwer Academic Publishers. A report from Russian scientists investigating radioactive pollution in the area of Semipalatinsk where the testing of nuclear weapons was carried out.

Lovelock, J. 1982. *Gaia: A new look at life on earth*. Oxford: Oxford University Press. A concise description of Gaia theory by its propounder.

Nosengo, N. 2005. Fertilised to death. *Nature* 425: 894–895. A short communication about eutrophication.

Osborn, A. 2011. How Chelyabinsk became synonymous with pollution. *Daily Telegraph*, July 27 2011. A readable account of radioactive pollution arising from the mismanagement of a nuclear power station.

United Nations Environment Program (UNEP). 2011. *Bridging the gap*. UNEP Synthesis report from a meeting in Nairobi, Kenya. A report on global warming.

Section II

Examples of Pollutants

The first section of this book described some basic principles underlying the harmful effects of pollutants upon the natural environment. Included here were the problems associated with persistent pollutants that can undergo biomagnification in food chains, the ecological significance of sublethal effects such as changes in behavior or reproductive success, the problem of secondary effects of pollutants (e.g., the consequences of removal of natural enemies), and the knock-on effects of disturbance of natural cycles. This section will focus on some examples of pollution to illustrate how these principles work out in practice.

Much of this section is devoted to man-made pollutants that were discovered during the latter part of the twentieth century. Over this period there were very rapid advances in the discovery of new industrial chemicals, including pesticides, together with a growing awareness of the side effects that they may have on ecosystems. Before dealing with these, however, there will be a brief consideration of earlier problems that arose with the rapid increase in numbers of the human race, and associated urbanization and industrialization. With the growth of towns, serious problems of water and air pollution arose—problems that were not restricted to the urban environment because both water and air are highly mobile.

8 Early Problems Connected with Urbanization and Mining

INTRODUCTION

During the early history of the human race there was no urbanization. The picture that comes through to us is one of cave dwellers and nomadic tribes. Early urbanization became evident some 1000 years BC with the emergence of Chinese, Egyptian, Assyrian, Persian, and Babylonian civilizations. Rather overlooked, there were also early civilizations in South America that left behind monumental architecture, e.g., that of Chavin de Huantar in Peru, dating back to c. 700 BC. As towns grew in size there were inevitably problems of waste disposal—not least of which was human waste.

Moving forward to the seventeenth century, an early problem of sizable towns was the disposal of raw sewage into rivers. In the late seventeenth century John Evelyn wrote a diary that records, among other things, problems of pollution. In nineteenth century Victorian London the Thames River was filthy, stinking, and without fish. As discussed in Chapter 4, organic waste has a high biological oxygen demand, and this can lead to the deoxygenation of water and the death of fish (Figure 4.4). In time, this problem came to be resolved, at least in part, by the treatment of wastewaters in sewage plants. Organic residues were broken down and treated sewage subsequently discharged into rivers. This practice was largely successful, but in some places, e.g., the Thames estuary in London, there remained something of a problem even into the 1950s. Oxygen levels remained low, and the worst affected reaches were unable to support fish (Mellanby 1967). With time, however, sewage treatment practices have been refined in Western countries, and the quality of freshwater of rivers has improved. This has been an important issue with seaside resorts that once discharged raw sewage into the sea, making it unhealthy for people to swim. Standards have become stricter in recent years in the European Union, where bathing beaches are graded according to the water quality of the sea. There has recently been considerable improvement in sewage treatment to ensure that seawater is clean enough for bathing.

The quality of surface waters, however, continues to be an issue. With industrial progress new contaminants have appeared in sewage effluents. Industrial chemicals, pesticides, and drugs in sewage have been the subject of interest and concern, and sometimes refinements of treatment procedures have been necessary to avoid the emergence of new pollution problems. Examples will be given in the following chapters. An older problem of sewage treatment, however, has been the widespread use of soaps and detergents, which will now be considered.

DETERGENTS IN SURFACE WATERS

Before the appearance of synthetic detergents, soaps were widely used for washing and cleansing. Soaps are sodium or potassium salts of long-chain fatty acids such as palmitic acid. Soaps are chemicals that have both polar and nonpolar characteristics. The polarity is due to the negative charge that they carry. The nonpolarity is due to the long fatty acid chain to which this charge is attached, as illustrated below:

$$CH_3(CH_2)_{14}COOH \rightarrow CH_3(CH_2)_{14}COO^- + H^+$$

Palmitic acid Palmitate ion

Sodium or potassium salts of palmitic acid release the palmitate ion when they are added to water. If greasy hands are washed in this soapy solution, the nonpolar fatty acid chains dissolve in the grease and the negative charge remains in the water, thus drawing the grease away from the hands and into the water. Because of this dualistic property soaps can be used for many cleaning purposes that require the removal of oily or greasy substances. Detergents (Figure 8.1) work by the same principle.

Soaps were early, readily biodegradable contaminants of surface waters. Detergents can be more effective cleaning agents than soaps and have come to be very widely used. As can be seen, they are of three basic kinds: (1) anionic detergents that resemble soaps in bearing a negative charge, (2) cationic detergents that carry a positive charge, and (3) nonionic detergents. The latter may seem to be a contradiction in terms. However, while they are not fully ionized, they do have small degrees of negative or positive polarity distributed over the molecule—a consequence of the electron-attracting power of the oxygen atoms contained within their structure.

At one time so-called hard detergents such as sodium tetrapropylene benzene sulfonate caused rather conspicuous environmental problems. Because of their molecular structure they were not so readily broken down during the course of sewage treatment as were soft detergents such as sodium alkyl sulfonate (see Figure 8.1). Their branched chain structure made it more difficult for bacterial enzymes to degrade them. Consequently, significant quantities of them found their way into surface waters, and when the water was aerated at weirs, masses of foam were formed, sometimes referred to as detergent swans. These swans sometimes moved out of rivers onto neighboring fields and roads (see Mellanby 1967).

There was no doubting the unsightliness of this phenomenon, but there was some debate about what ecological or other damage this caused. Fishermen claimed success when casting their lines into foamy waters. Indeed, it was questionable whether, according to the definition used in this account, the detergents should be regarded as pollutants or merely contaminants. However, it was clear that detergents were retarding the uptake of oxygen in waters where there was already a shortage of it. They

Anionic

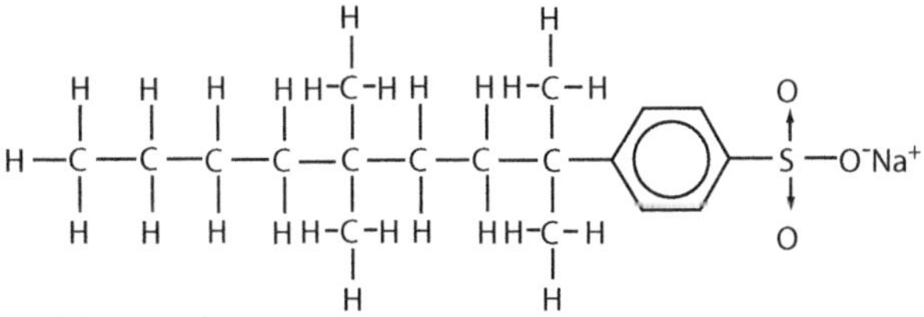

Sodium tetrapropylene benzene sulphonate (hard)

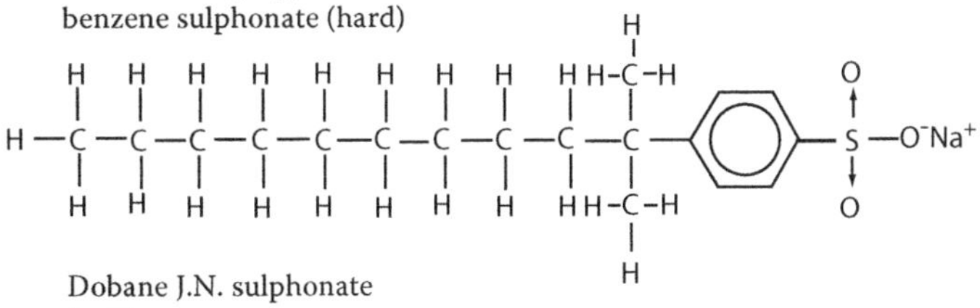

Dobane J.N. sulphonate (soft)

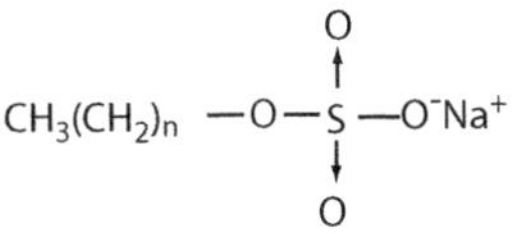

$CH_3(CH_2)_n$ —O—S—O^-Na^+ Sodium alkyl sulphonate (soft)

Cationic

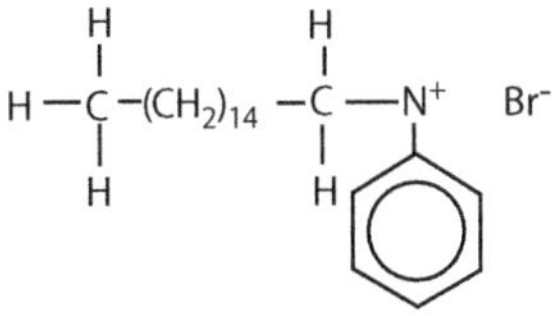

$H-C-(CH_2)_{14}-C-N^+$ Br^-

Cetyl pyridinum bromide

Nonionic

Polyglycol ethers of alkylated phenols, e.g., lissapol N stergene

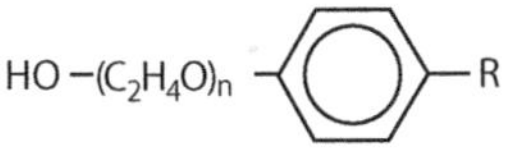

$HO-(C_2H_4O)_n$ —R R = alkyl group

FIGURE 8.1 Detergent molecules contain both polar and nonpolar elements. They may have permanent negative charges (anionic detergents), permanent positive charges (cationic detergents), or a collection of small positive and negative charges over their structures (nonionic detergents).

were also making the purification of the sewage less efficient. In due course they came to be replaced by more readily biodegradable soft detergents.

AIR POLLUTION

As explained earlier, air pollution has been a natural phenomenon during the course of evolutionary history, for a very long period prior to the emergence of the human race. Sulfur oxides have been emitted by volcanoes and aromatic hydrocarbons (e.g., polycyclic aromatic hydrocarbons (PAHs)) generated by forest fires. With the growth of urbanization and the development of industry these and a variety of other chemicals have been released into the atmosphere as a consequence of the burning of fossil fuels, mining, and the smelting of ores and emissions from chemical factories.

During the industrial revolution towns became affected by smoke and soot originating from the burning of fossil fuels. This development was well documented by Charles Dickens in the mid-nineteenth century with his descriptions of London and northern industrial towns at this time. In fact, warning bells had been sounded in the previous century when Sir Percival Pott related scrotal cancer in chimney sweeps to exposure to soot and coal tar. It is now realized that soot, coal tar, and other products of incomplete combustion of fossil fuels contain polycyclic aromatic hydrocarbons (PAHs) such as benzo(a)pyrene, some of which are potent carcinogens. More recently the work of Richard Doll and others has implicated compounds of this kind in the development of lung cancer by heavy cigarette smokers. Towns could be very unhealthy places in which to live during this period. Apart from the problem of lung cancer, polluted air has been shown to cause serious respiratory problems such as bronchitis in vulnerable people, for example, in the London smogs of the years following the World War II.

During the latter part of the twentieth century, the internal combustion engine made a significant contribution to air pollution in densely populated areas. At one time, significant levels of oxides of nitrogen, carbon monoxide, hydrocarbons, and lead (both inorganic and organic) were released from exhausts. More recently, strict controls have been introduced of emissions from internal combustion enzymes. Catalytic converters are now fitted to engines to prevent undesirable levels of pollutants in exhaust emissions.

One widely publicized example of air pollution in towns was the photochemical smog that became a problem in the Los Angeles basin in the mid-twentieth century. This developed under clear skies and arose from a complex series of photochemical reactions between nitrogen oxides and hydrocarbons released by vehicles, and was driven by solar energy. One of the products of these reactions was peroxyacetyl nitrate, which acts as an eye irritant.

Lead was also recognized as a serious pollutant during these postwar years. The presence of lead in car exhausts was due to the practice of adding a mixture of lead tetraalkyls to petrol, to act as an antiknock. Both tetraalkyl lead and inorganic lead were released in car exhausts (Bryce-Smith 1971). More recently, there have been restrictions on lead additives to petrol in many countries because of the health risk posed to humans.

Historically, much of the concern about air pollution centered on human health hazards in urban areas. Very little consideration was given to effects on other species living in urban environments. Neither was much thought given to effects on the natural environment in rural areas adjoining towns—despite the fact that air, like water, is very mobile, and pollution originating in towns will find its way into the countryside due to transportation in air and in rivers. Things have changed in more recent times, and there is now greater awareness of the damage that urban air pollution can cause to the natural environment.

The phenomenon of acid rain was discussed earlier (Chapter 7). A related problem has been the phytotoxic action of SO_2 in forestlands. In industrial areas of North Bohemia and Silesia (Czech Republic and Poland) much brown coal has been used, and this fuel has a high sulfur content. The large-scale burning of it in these areas led to the release of large quantities of SO_2 into the atmosphere. During communist times this practice caused extensive damage to pine forests (see introduction in Peakall et al. 1997). Since then emissions have been greatly reduced and there has been some recovery of the conifer forests.

MINES AND SMELTING

Mining for metals has been going on for a long time. Old mining sites still show high levels of metals such as copper, tin, lead, zinc, etc., in their soil, often associated with sparse vegetational cover. As we have seen in Chapter 5, such sites sometimes have strains of plants that show resistance (tolerance) to metals present at high levels in the soil (Chapter 5), bearing testimony to the toxic effects of these metals in earlier times. One such example is the site of an old copper mine at Parys Mountain in Anglesey, Wales. During the nineteenth century this was the largest copper mine in the world. It closed around the end of that century, but recolonization by vegetation has been slow on account of the high levels of copper still in the soil.

Smelting works as well as mining sites have been heavily contaminated with metals. One of the most studied examples is that of a smelting works for lead, cadmium, and zinc located at Avonmouth near Bristol, England (Hopkin 1989). In the close vicinity of the factory concentrations of lead, zinc, and cadmium have been found that are at least two orders of magnitude higher than normal background levels. In heavily contaminated areas there are no earthworms, woodlice, or millipedes. Dead vegetation has accumulated on the surface as a thick layer, because of the absence of organisms that would normally fragment leaf litter.

SUMMARY

With the growth of towns and the development of mining and industry there came problems of air pollution and water pollution in urban areas. The combustion of coal led to the release of soot and smoke, which contained carcinogenic pollutants. The disposal of human wastes became a problem and led to the release of large quantities of untreated sewage into rivers. Towns became unhealthy places in which to live. In time the position improved with control of emissions from chimneys and the treatment of sewage before release into water courses.

Into the twentieth century other pollution problems emerged. Hard detergents that were not effectively broken down by sewage treatment found their way into rivers and caused pollution problems. Strikingly obtrusive were detergent swans that formed at weirs, and sometimes came inland. The introduction of soft biodegradable detergents to replace hard detergents overcame this problem. With the great increase in the number of motorcars, there came a further problem—the release of lead compounds from car exhausts, which was seen to constitute a human health hazard. This issue came to be resolved by the introduction of strict controls in the use of lead compounds as additives to petrol.

In Eastern Europe after the World War II industrialization led to serious pollution problems. One was the burning of brown coal in the areas of Silesia and North

Bohemia. This caused serious pollution by sulfur dioxide that was responsible for both urban pollution and extensive damage to forests. Since the fall of communism this has been brought under control.

To a large extent the main concern about these cases of pollution has been the human health hazards that they have presented. However, pollutants can be readily moved by rivers and the movements of air, and damage to the natural environment is also an issue.

FURTHER READING

Hopkin, S.P. 1989. *Ecophysiology of metals in terrestrial invertebrates.* Barking, UK: Elsevier Science. A useful textbook dealing with the effects of metals on soil organisms in polluted areas.

Mellanby, K. 1967. *Pesticides and pollution.* London: Collins. Gives a readable historical account of urban pollution and contains some pertinent illustrations.

Peakall, D.B., Walker, C.H., and Migula, P. 1997. *Biomarkers: A pragmatic basis for remediation of severe pollution in Eastern Europe.* Dordrecht: Kluwer Academic Publishers. Features a number of examples of pollution problems that arose in Eastern Europe during communist times.

9 The Organochlorine Insecticides

INTRODUCTION

The first of the organochlorine insecticides to be widely used was DDT. Discovered by Paul Muller of the Swiss firm Geigy in 1939, it was used by Allied forces during World War II for vector control. After the war it quickly gained ground as an insecticide for many purposes—for controlling agricultural and horticultural pests, vectors of disease, houseflies, cockroaches, and other nuisance organisms, and parasites of farm animals. It was some time before the downside of this wonder insecticide came to be recognized. Rachel Carson's book *Silent Spring*, published in 1962, drew attention to undesirable side effects it was having in the natural environment and raised questions about the possible human health hazards that might be presented by residues of DDT.

Some time later, into the 1950s, a group of organochlorine insecticides called the cyclodienes were introduced in North America and Western Europe. These included the compounds aldrin, dieldrin, and heptachlor. These were, in many cases, more toxic to vertebrates than was DDT, and residues of them were found to be highly persistent in animals and in soil.

In the following account the chemical and biological properties of (1) DDT and related compounds and (2) the cyclodienes will be reviewed separately before considering their environmental side effects collectively. Declines of populations have been studied when both DDT and cyclodienes were present in the environment—raising problems of distinguishing between the effects of one type of insecticide and the other. The account that follows will attempt to distinguish between the effects of DDT residues and those of cyclodiene residues. There will also be a brief mention of some other organochlorine pesticides that have been of lesser environmental concern.

DDT AND RELATED PESTICIDES

The active ingredient of commercial DDT is its p,p′ isomer (see Figure 9.1). p,p′-DDT is a stable crystalline solid that is persistent in soils, surface waters, and food chains. It is stable in both a chemical and a biochemical sense. It is fat soluble and has a very low solubility in water. It is not rapidly broken down in the natural environment by either chemical or biochemical processes and is very persistent.

The metabolic transformations shown in Figure 9.1 all take place slowly. A particularly significant one is the transformation into the highly persistent metabolite p,p′-DDE. In the longer term, over months and years, this is the principal DDT metabolite remaining in living organisms and in soils and surface waters after the

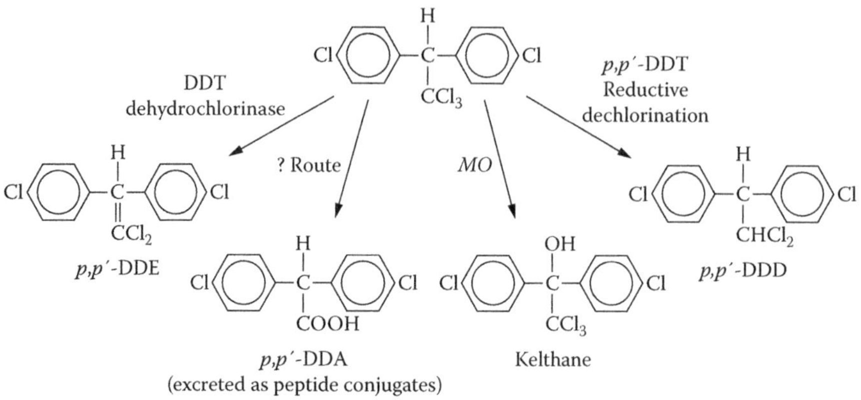

DDT dehydrochlorinase

? Route

MO

p,p′-DDT Reductive dechlorination

CCl_3

p,p′-DDE

CCl_2

p,p′-DDA (excreted as peptide conjugates)

COOH

Kelthane

CCl_3

p,p′-DDD

$CHCl_2$

FIGURE 9.1 The metabolism of p,p′-DDT.

insecticide has been used. It is considerably more persistent than DDT itself. The sublethal effects of this metabolite will be discussed late. Another metabolite is the compound p,p′-DDD. This is formed by anaerobic metabolism in dead animal tissues and by anaerobic microorganisms. It has been used as an insecticide in its own right under the commercial name of rhothane. The water-soluble metabolite p,p′-DDA is slowly formed by animals and is readily excreted. The oxidative metabolite kelthane has been marketed as an acaricide.

p,p′-DDT and some of its relatives act as nerve poisons. They are able to bind tightly to a site located on certain sodium channels that pass through the nerve membrane (Figure 9.2). These channels have a vital role in regulating the passage of messages along the nerve. When a message is sent along a nerve, it takes the form of an electrical impulse that moves very rapidly, usually only for a very short period of time. This impulse (action potential) is initiated by a flow of sodium ions through the channel, and is normally rapidly terminated. When a nerve is affected by DDT, the pore stays open for too long and the flow of ions is unduly prolonged, with the consequence that the passage of the action potential along the nerve is disrupted.

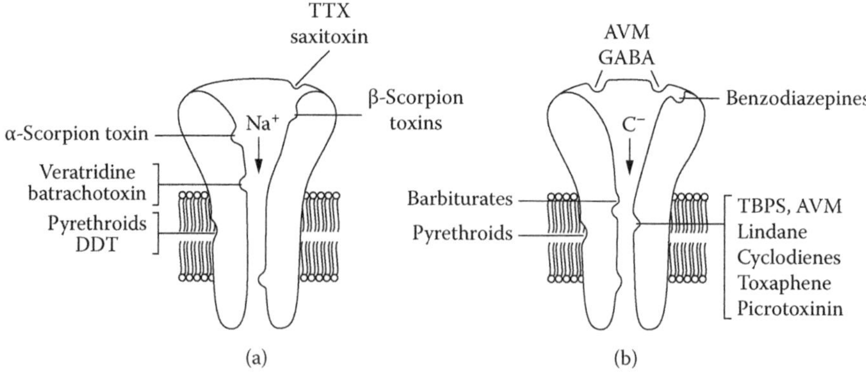

FIGURE 9.2 Sites of action of organochlorine insecticides.

Outwardly, this effect may show itself in the form of uncoordinated tremors or twitches—symptoms of poisoning by DDT. The acute oral lethal toxicity of DDT is 100–2500 mg/kg to mammals and >500 mg/kg to birds. It is distinctly less toxic to these organisms than dieldrin.

The persistent metabolite, p,p′-DDE, is able to cause the thinning of avian eggshells. Basically, this is because it can inhibit the transport of calcium ions (Ca^{2+}) from blood to the developing eggshell contained within the shell gland of birds. There is still some uncertainty about the biochemical mechanism involved. There is some evidence that p,p′-DDE has an inhibitory effect upon an enzyme called calcium ATP-ase, which has a role in transporting Ca^{2+} from blood into the eggshell gland. This action of p,p′-DDE will be discussed further later in this chapter when considering its effects upon populations of predatory birds.

THE ENVIRONMENTAL FATE OF P,P′-DDT AND RELATED COMPOUNDS.

Some data on the persistence of p,p′-DDT and p,p′-DDE are given in Table 9.1. p,p′-DDT itself and its metabolite p,p′-DDD are highly persistent in soils, with half-lives of 2.8 and 10+ years, respectively, in the examples given. The even more persistent metabolite p,p′-DDE has estimated half-lives in soil as high as fifty years. These values are representative of temperate soils. Both compounds are less persistent in animals where they are broken down into metabolites by enzymic action, as shown in Figure 9.3. Nevertheless, these are still high degrees of persistence. The relatively long half-life of p,p′-DDE in animals explains its tendency to undergo strong biomagnification with movement along food chains.

The half-lives of these two compounds in soil vary according to soil type. Persistence is greatest in heavy soils that are high in clay minerals and organic matter. Persistence is less in tropical soils than in temperate ones. This is mainly because they have appreciable volatility—and will tend to vaporize faster at the high temperatures of tropical soils than they will from cooler temperate soils.

TABLE 9.1
Half-Lives of DDT and Related Compounds

Chemical	Material/Organism	Half-Life (t_{50})
p,p′-DDT	Soil	2.8 years
p,p′-DDD	Soil	10+ years
p,p′-DDT	Feral pigeon	28 days
p,p′-DDE	Feral pigeon	250 days
p,p′-DDT	Rat	57–107 days
p,p′-DDT	Chicken	36–56 days

Source: Walker (2009). *Organic Pollutants: an Ecotoxicological perspective*, 2nd ed.

Note: For explanation of half-life, see Box 9.1.

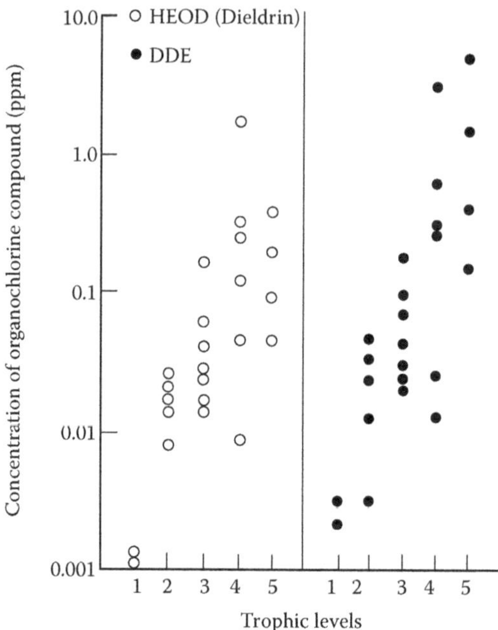

FIGURE 9.3 Organochlorine residues in the Farne Island ecosystem. (From Walker, C.H., *Environmental Pollution by Chemicals*, 2nd ed., London: Hutchinson, 1975. With permission.)

At one time DDT was widely used as an insecticide on agricultural crops and in orchards to control insect pests. The persistence of these compounds in agricultural soils was sometimes found to give rise to unacceptably high residues in crops grown in them, an observation that led to the placing of restrictions and bans on the use of DDT. The extensive use of DDT in the United States, Canada, and Western Europe resulted in the appearance of undesirably high levels of DDE at the top of both terrestrial and aquatic food chains. This problem will be discussed in more detail later in this chapter.

p,p′-DDT and certain of its metabolites can undergo biomagnifications in both aquatic and terrestrial food chains. One of the earliest indications of this came from a study of the residues of p,p′-DDD in the ecosystem of Clear Lake, California (Hunt and Bischoff 1960). During the 1950s applications of the insecticide rhothane, whose active ingredient is DDD, were made to the lake. The aim was to control a gnat, whose larvae inhabit water, and create a nuisance for visitors to the lake. In time, it was noticed that a piscivorous bird, the western grebe, declined in numbers. Some of these birds were found dead and taken for analysis. The high levels of p,p′-DDD found in the body fat of birds raised suspicions that they were dying from DDD poisoning. A study of residues in samples taken from this ecosystem suggested strong biomagnification with movement of p,p′-DDD along the food chain. The levels of this compound in biota expressed in ppm by weight were as follows: plankton (5.0), nonpredatory fish (40–100), predatory fish (80–1000), and western grebes (1600). The levels in fish and grebes were measured in depot fat.

BOX 9.1 HALF-LIVES

A half-life is the time that it takes for the concentration of a residual chemical to halve. In ecotoxicology half-lives are measured in both biota (animal or plant tissues) and the abiotic environment (e.g., soils or surface waters). In the simplest case the rate of loss is proportional to the logarithm of the concentration of the chemical over the period in question—as happens in the radioactive decay of radioisotopes. This is illustrated by the linear graph shown in Figure 9.4. Such steady exponential loss may be shown by persistent chemicals present in animals, soils, or lakes during certain stages of their elimination, but not necessarily over the entire period.

Not infrequently the picture is more complex. For example, an initially rapid but decreasing rate of loss may be succeeded by a slower and more constant phase of exponential loss. This is illustrated by data for loss of persistent organochlorine insecticides from soil shown in Figure 9.5. In this changing situation it is not possible to calculate one half-life for elimination over the entire period. During the early stage there is relatively rapid loss, much of it due to volatilization of the pesticides. However, in the longer term persistent pesticides become strongly bound to soil colloids (clay and organic matter) and are only slowly lost due to volatilization or biodegradation. During this later stage there is steady exponential loss, and it is possible to calculate a half-life. Over the entire period it is possible to calculate a period during which, for example, 95% loss occurs, albeit not at a constant rate. For further discussion see Edwards (1976).

The strong biomagnification of p,p′-DDE that can occur in marine food chains was illustrated by the results of a study of organochlorine residues in the ecosystem of the Farne Islands, off the east coast of England, which was conducted in the mid-1960s (Robinson et al. 1967) (Figure 9.3). Five trophic levels were identified in this ecosystem. The lowest level included aquatic plants such as brown algae. Aquatic invertebrates and fish were represented in levels 2 and 3, and vertebrate predators occupied levels 4 and 5. Groups 4 and 5 included predatory fish and piscivorous seabirds such as cormorants, shags, and members of the auk family.

Residues of p,p′-DDE and dieldrin in species from different trophic levels are plotted on a log scale. It can be seen that residues of p,p′-DDE were some 1000-fold higher in predators at the top of the food chain than in the organisms of trophic level 1. Piscivorous birds in trophic level 5 had c. 50 times higher concentrations of this residue than did the fish upon which they were feeding (e.g., sand eels); these fish belonged to trophic levels 2 and 3.

There has also been evidence of marked biomagnifications of p,p′-DDE in terrestrial food chains. DDT was once widely used for control of pests in orchards. In one study conducted in an orchard in Norfolk during the period 1971–1972 (Bailey et al. 1974; Walker 2009) the following levels of p,p′-DDE were found, expressed in ppm by weight, in organisms from different trophic levels:

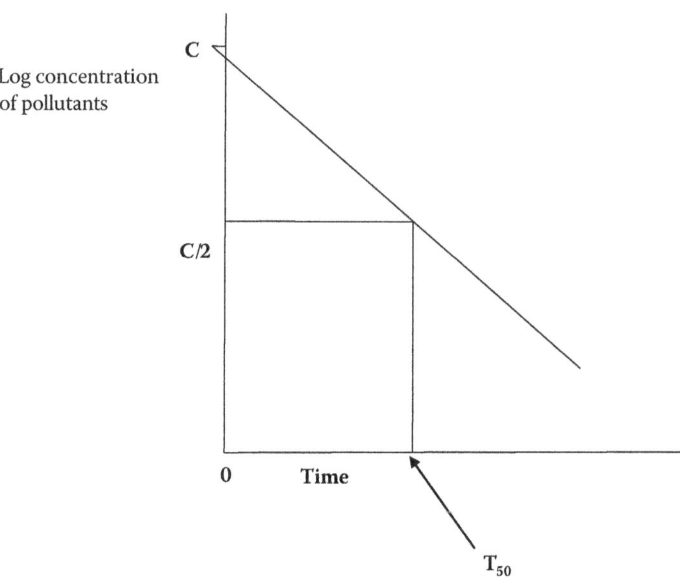

FIGURE 9.4 Estimating half-life.

Soil	0.5–1.1
Earthworms	1.4–4.2
Blackbirds and song thrushes	24–192

Blackbirds and song thrushes (*Turdus* spp.) feed upon earthworms, slugs, and snails in orchards, and these results suggest that there may have been marked biomagnification of this compound by the birds from their food—perhaps by as much as 20-fold. These members of the thrush family are preyed upon by sparrowhawks, a species that also acquired high residues of p,p′-DDE at this time in areas of Sussex and Kent where there were many apple orchards.

This tendency of p,p′-DDE to build up in terrestrial and aquatic food chains was also indicated in a survey of organochlorine residues in different species of British birds and their eggs during the early 1960s (Moore and Walker 1964) (Figure 9.6).

The highest residue levels of p,p′-DDE were found in predatory birds. In breast muscle samples, the highest levels were in sparrowhawks and barn owls (terrestrial habitat) and in herons and great crested grebes (aquatic environment). Considerably lower levels were found in herbivores and omnivores in both types of habitat. In samples of birds' eggs, the same trend was found. Peregrines and sparrowhawks (terrestrial habitat) and herons and great crested grebes (aquatic habitat) had much higher levels than did pheasants and moorhens. The latter two species are largely herbivorous, and from terrestrial and aquatic habitats, respectively.

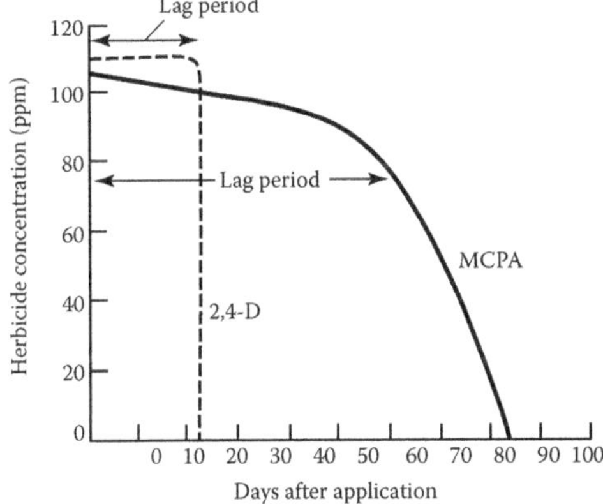

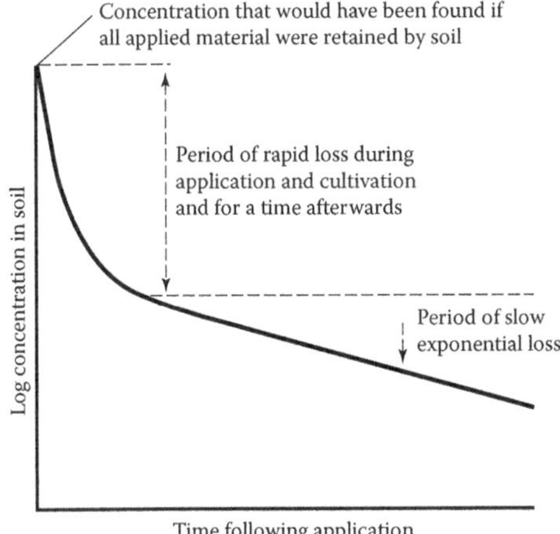

FIGURE 9.5 Loss of persistent organochlorine insecticides from soils.

THE CYCLODIENE INSECTICIDES

The cyclodiene insecticides were first introduced into Western Europe and North America in the early 1950s. However, they were not used on a large scale until the mid-1950s. The most important of these from an environmental point of view have been aldrin, dieldrin, and to a lesser extent, heptachlor. Other examples are chlordane, endrin, and endosulfan. Like DDT, these compounds are stable solids, with low water solubility but high lipid solubility. The structures and metabolism of aldrin and dieldrin are given in Figure 9.7.

Aldrin and dieldrin have chlorinated cage structures (Figure 9.7). Epoxidation of aldrin converts it into dieldrin and occurs fairly rapidly within many animals.

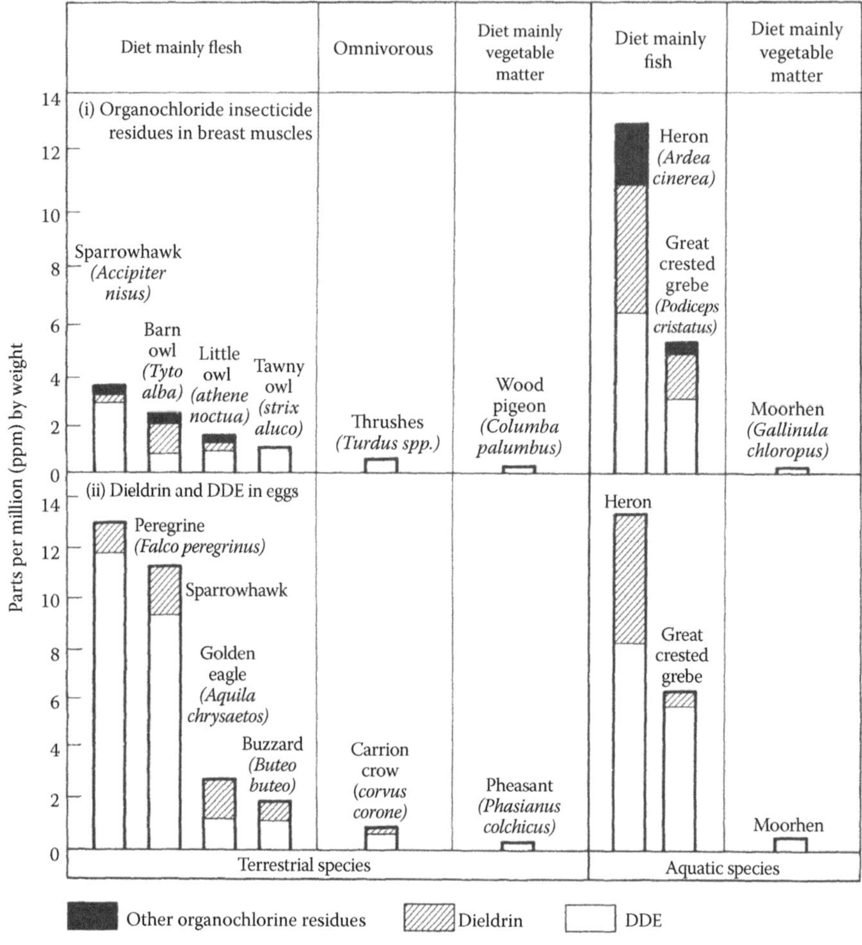

FIGURE 9.6 Organochlorine insecticide residues in British birds and their eggs during the early 1960s. (From Walker, C.H., et al., *Principles of Ecotoxicology*, 4th ed., Boca Raton, FL: Taylor & Francis/CRC Press, 2012.)

Consequently, aldrin is rarely found in significant quantities as a residue in animals that have been exposed to it in the field. The residue found is largely or entirely dieldrin. Heptachlor has similar structure and properties to aldrin, and is epoxidized in animals to heptachlor epoxide, which has similar properties to dieldrin. The principal residue of heptachlor in exposed animals is heptachlor epoxide.

The main routes of metabolism of dieldrin (shown in Figure 9.7) involve oxidation or hydration. The two metabolites containing hydroxyl groups have some water solubility and are readily excreted. All very well, but the main problem is that they are only formed very slowly, and in consequence, dieldrin has a long half-life in most animals, notably in man (Table 9.2). Aldrin has a much shorter half-life than dieldrin

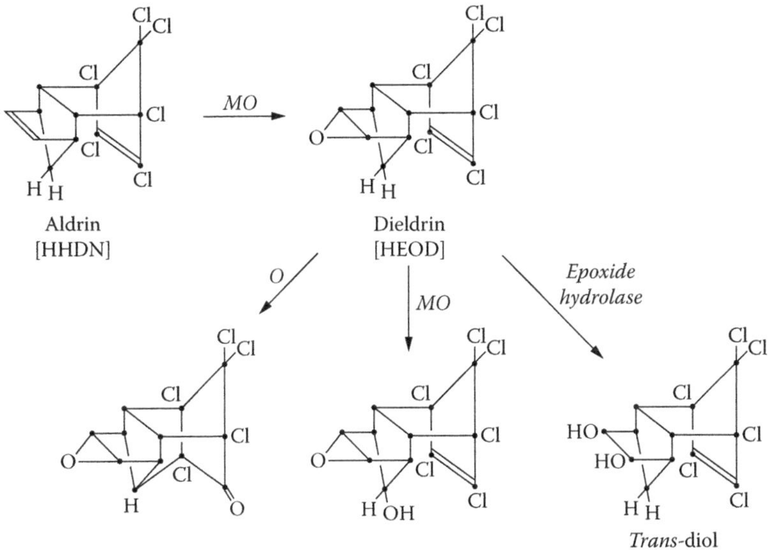

FIGURE 9.7 Metabolism of some cyclodienes.

in soils and, as we have already seen, a very short one in vertebrates. An important factor in determining the slow metabolism of dieldrin and p,p′-DDT is the high level of chlorination. The bulky chlorine atom acts as an obstacle to enzymic degradation. Other highly chlorinated molecules, such as PCBs and dioxins, are also very persistent for the same reason (Chapter 13). The same consideration also applies to halogen elements other than chlorine, e.g., bromine and iodine. Substitution of them on organic molecules can retard metabolic attack by enzymes.

Dieldrin, like p,p′-DDE, is liable to undergo strong biomagnification with movement along food chains (see, for example, Figures 9.3 and 9.6). Both of these two very persistent organochlorine residues had harmful side effects upon predatory vertebrates at the top of aquatic and terrestrial food pyramids when DDT and cyclodiene insecticides were widely used.

Dieldrin, like DDT, acts as a nerve poison, but it does so at a different site of action and in a different way. This distinction is seldom made in popular articles on the subject, where all organochlorine insecticides tend to be lumped together and effects of dieldrin are liable to be attributed to DDT. Dieldrin and heptachlor epoxide interact with GABA receptors, which are located in the vicinity of inhibitory synapses of the nervous system in both vertebrates and invertebrates (Figure 9.2). These receptors are located on pore channels that pass through the nerve membrane. The term *GABA* is an acronym for the neurotransmitter gamma amino butyric acid, which interacts with these sites and thereby regulates the passage of chloride ions (Cl^-) through them. Binding of dieldrin to this site causes disturbances in the transmission of nervous impulses in the region of inhibitory synapses. GABA receptors are located in the brain of vertebrates and invertebrates; they are typically also found in the peripheral nervous system. When vertebrates

TABLE 9.2
Half-Lives of Dieldrin and Aldrin

Compound	Material/Organism	Half-Life (t_{50})
Dieldrin	Soil	2.5 years
Aldrin	Soil	0.3 years
Dieldrin	Male rat	12–15 days
Dieldrin	Pigeon	47 days (mean)
Dieldrin	Man	369 days (mean)

Source: Data from Edwards, C.A. (1973). *Persistent Pesticides in the Environment*, 2nd Edition. CRC Press. Robinson, J., et al., *Nature* 214: 1307–1311, 1967; Environmental Health Criteria 91. Aldrin and Dieldrin. Geneva. WHO 1989.

are poisoned with dieldrin, they often show convulsions. Birds poisoned with dieldrin often show clenched claws. At sublethal levels, dieldrin can cause a range of behavioral effects upon animals (including humans). Symptoms include hyperirritability, dizziness, drowsiness, and nausea.

The acute oral lethal toxicity of dieldrin to mammals or birds lies in the range of 30–100 mg/kg. It is considerably more toxic than DDT to most vertebrates.

THE ENVIRONMENTAL FATE OF CYCLODIENES

The following account will be restricted to dieldrin, which has been of particular concern to environmentalists. Aldrin, another hazardous cyclodiene insecticide, is rapidly converted to dieldrin, and very little has been found as a residue in biota exposed to it. Thus, an unknown proportion of dieldrin detected in free-living organisms originated as aldrin. Heptachlor has been less widely used than either aldrin or dieldrin, and its principal residue, heptachlor epoxide, has similar properties to dieldrin. In some interpretations of the likely effects of cyclodiene residues in vertebrates, dieldrin and heptachlor epoxide have been lumped together because they have the same mode of action, similar properties, and similar toxicity to vertebrates.

In the studies of persistent organochlorine insecticide residues in biota performed in Britain during the early 1950s dieldrin, as well as p,p′-DDE, was found to undergo biomagnification with movement along food chains, the highest concentrations occurring in animals of the highest trophic levels (Figures 9.3 and 9.6). In the Farne Islands ecosystem p,p′-DDE was, in general, biomagnified to a greater extent than dieldrin. Some predators in trophic level 5 contained residues over 1000-fold higher than existed in organisms of trophic level 1. Piscivorous birds such as the shag had residues some 50-fold higher than those in the fish upon which they were feeding. This suggests a high degree of bioaccumulation, as these fish represent the principal source of p,p′-DDE to the bird. Dieldrin

BOX 9.2 THE METABOLIC FACTOR IN BIOMAGNIFICATION

Dieldrin and p,p′-DDE are biomagnified to a striking extent with progression along food chains—both terrestrial and aquatic. Higher chlorinated PCBs and dioxins show a similar tendency. Why is this? Referring to Figure 2.1, it can be seen that metabolism to water-soluble and readily excretable products is a central factor in the elimination of foreign lipophilic molecules from the bodies of animals. Fat-soluble molecules tend to be retained in fatty locations in the body unless they are converted into water-soluble products that can be readily excreted. It follows that loss of such molecules from the body tends to be relatively slow if metabolism is slow. Conversely, loss tends to be relatively rapid if metabolism is fast.

With terrestrial animals this relationship between rate of metabolism and rate of excretion is fairly straightforward—because they have little ability to excrete highly lipophilic compounds directly. By contrast, fish are able to lose these compounds by diffusion across gills into the ambient water. A similar situation exists with aquatic invertebrates. In these aquatic species enzymes such as P450-based monooxygenases that can convert lipophilic compounds into water-soluble products are poorly represented. It can be argued that, in contrast to terrestrial species, there has not been strong evolutionary pressure for the production of such enzymes.

A study of the activity of monooxygenases toward lipophilic xenobiotics such as certain organochlorine pesticides and drugs revealed marked species differences (Walker 1978, 1980). Terrestrial vertebrates had relatively high enzyme activities, whereas fish had relatively low enzyme activities. Raptors and piscivorous birds had lower activities than herbivorous or omnivorous birds. It was suggested that these differences could be explained by differential evolutionary pressure. As part of plant animal warfare, herbivores and omnivores developed enzyme systems that could degrade lipophilic plant toxins. Predators, on the other hand, were not challenged by plant toxins to any important extent, and so did no need to develop metabolic systems to promote their excretion.

showed an apparent *biomagnification* factor of some 200- to 300-fold if comparing members of trophic level 5 with occupants of trophic level 1. Interestingly, the piscivorous shag had an apparent bioaccumulation factor of 63-fold when comparing the dieldrin level in it with that in its principal prey, the sand eel (Robinson et al. 1967).

EFFECTS OF DDT AND CYCLODIENES ON BIRDS OF PREY

As we have seen, organochlorine insecticides began to come into use in Western countries shortly after World War II. DDT was the first of these to be widely used. The cyclodienes appeared in the early 1950s, and came into widespread use in some

TABLE 9.3
Raptors and Pesticides in the UK: Chronology

Observations	Year(s)	Pesticide Usage
Eggshells of sparrowhawks and peregrines became markedly thinner in many areas of the UK	1947–1949	DDT came into widespread use
Evidence of some localized declines of sparrowhawk populations in intensely arable areas of southeast England	1950–1954	
Large kills of grain-eating birds on agricultural land caused by dieldrin poisoning; also secondary poisoning of predators including raptors; population crashes of sparrowhawk and peregrine in or near areas where dieldrin, etc., seed dressings were used; remote areas (e.g., Scottish Highlands) relatively unaffected	1955–1962	Dieldrin and related cyclodiene insecticides came into widespread use notably as cereal seed dressings
Partial recovery of sparrowhawk and peregrine populations in affected areas; sparrowhawks, however, remained close to extinction in most arable areas of east England (Lincolnshire, Cambridgeshire, Bedfordshire)	1963–1974	A series of bans/restrictions placed upon the use of organochlorine insecticides in agriculture, including upon dieldrin, etc., seed dressings on spring-sown cereals
Recovery of peregrine continued	1975	Ban on use of dieldrin seed dressings on autumn-sown cereal
Sparrowhawks progressively recolonized worst affected eastern areas and bred successfully—despite tissue DDE levels high enough to cause eggshell thinning	1980–1990	

Western countries around the middle of the decade. The cyclodienes were extensively used, notably as seed dressings, in the United Kingdom, the Netherlands, and Denmark. The timetable for these events in the UK is shown in Table 9.3.

We have already seen that dieldrin and p,p′-DDE were both found to exist at relatively high levels in predatory birds in the UK during the mid-1960s. Also there was early evidence that both residues reached high enough concentrations to have harmful effects upon predatory birds. This discovery came at a time when there was compelling evidence for serious declines of raptors such as the sparrowhawk, the peregrine, and the merlin over many areas of the country. One advantage of Great Britain being a rather crowded island is that there have long been many ornithologists operating in a relatively small area, making it possible to conduct integrated countrywide surveys of bird populations. Such surveys were conducted during the period in question, making it possible to compare the residues of persistent organochlorine insecticides in raptors with population data from the field, to see whether the declines could be related to the levels of these residues. Support for these activities came from the British Trust for Ornithology (BTO) and the Royal Society for the Protection of Birds (RSPB). In what follows, the results of these countrywide studies in the UK will be described, following the timetable given in Table 9.3, before considering evidence from elsewhere addressing the same question.

DDT came rapidly into widespread use in Western countries shortly after World War II, and residues of its highly persistent metabolite p,p′-DDE became widely distributed here during this period. In 1967 Ratcliffe published a groundbreaking paper showing that the eggshells of British peregrines and sparrowhawks became thinner during the period in question (1946–1947). He speculated that this was related to environmental pollution by DDT (see Ratcliffe 1993). This was very controversial at the time, but evidence supporting his suggestion gradually grew, and a critical paper published in 1970 established that environmentally realistic levels of p,p′-DDE caused the thinning of egg shells when included in the food of American kestrels (Wiemeyer and Porter 1970). In due course, much more evidence from the laboratory and the field came forward to confirm this effect.

The average level of eggshell thinning in sparrowhawks and peregrines in the UK during the late 1940s varied considerably between individuals, showing a dependence on area. The majority of individuals showed thinning in the range of 14–18%, but in some cases, where DDT was widely used, values were higher than this (Ratcliffe 1967; Peakall 1993). Despite these findings, there was no evidence of any overall decline of either species at this time. Indeed, numbers of peregrines were increasing in coastal counties of England where they had been reduced to virtual extinction by shooting during the wartime. This was done to protect carrier pigeons bearing messages to and from occupied France. Sparrowhawks showed the greatest levels of eggshell thinning in some intensely arable areas of southeast England, where DDT was heavily used, but even here only a limited, localized decline in numbers was found around 1950 (Newton 1986). A global view of the effects of DDE-induced eggshell thinning on populations of predatory birds will be the subject of the next section.

The introduction of cyclodienes into Britain had dramatic consequences. From c. 1955 large-scale kills of wood pigeons, pheasants, and other granivorous birds occurred on agricultural land following the consumption of grain dressed with dieldrin and other cyclodiene insecticides. These neurotoxic insecticides caused tremors and convulsions and effectively immobilized poisoned individuals. At this time dressed cereal seed carried dieldrin residues of around 1000 ppm (i.e., c. 1000 mg/kg), and birds feeding on it were able to acquire lethal doses; c. 100 g of such dressed seed contained c. 100 mg of dieldrin. Thus, if a wood pigeon weighing 500 g consumed this quantity of dressed grain, the received dose would be c. 200 mg/kg. The acute oral median lethal dose (LD_{50}) for dieldrin to pigeons is c. 70 mg/kg.

There were also deaths of predatory birds such as sparrowhawks and kestrels feeding on prey heavily contaminated with dieldrin, which were consequently subject to secondary cyclodiene poisoning. Poisoned granivorous birds were "easy game" for these and other predators (e.g., foxes) because poisoning caused tremors and convulsions. These effects (1) drew the attention of predators and (2) effectively immobilized badly affected individuals.

Coincident with these events severe declines of peregrines and sparrowhawks began that were related in both time and space to the introduction of the cyclodiene insecticides. It should be reemphasized that these declines commenced eight years or more after the initiation of DDE-related eggshell thinning. Thus, contrary to some reports in the media, they were not attributable to the action of DDT. Adult sparrowhawks, for instance, only have a life expectancy of two–three years, so any population effects caused by eggshell thinning would have clearly shown years before these severe declines.

Taking the peregrine first (Ratcliffe 1993), population declines were most pronounced near arable areas where cyclodienes were widely used. The populations showing the least evidence of decline were in the mountainous areas of Scotland, far from arable farming. By the time that the problem was recognized, the peregrine had disappeared from the areas that were most badly affected, e.g., from most of the south coast of England. Consequently, there were very few specimens of birds or eggs from the worst affected areas that could be analyzed for organochlorine residues. Much work had to be done retrospectively. However, from the limited number of analyses carried out on liver and other tissues taken from birds found dead in the field, a substantial number contained sufficiently high levels of dieldrin to have caused lethal or sublethal effects. The question of residues and their interpretation will be returned to later in this chapter.

Unlike the peregrine, the sparrowhawk was fairly common in intensively farmed lowland areas before the introduction of cyclodienes. When surveys were conducted of sparrowhawks after the population crash of the later 1950s, they were found to have become virtually extinct in some lowland areas of eastern and southern England, where they had once been common (Newton 1986) (see Figure 9.8). On the other hand, they held their ground better in areas farther north and west. These are upland areas where stock farming is more important than intensive arable farming where dieldrin seed dressings were not much used. An interesting exception to this general trend was the area of the New Forest. The New Forest near the south coast has been kept as a national park for many years. There is little arable land here, and cyclodienes were not used to any significant extent in this area at the time of the population crashes of predatory birds. Sparrowhawks continued to breed successfully in the New Forest during the period in question.

During the early 1960s the problems associated with persistent organochlorine insecticides began to be recognized and research was undertaken on residues and population declines. In time, a series of bans and restrictions were placed on the marketing of cyclodienes and DDT that took effect over the period 1966–1976. Some population studies continued, and the peregrine continuously recovered from its low point of the early 1960s until 1991 (Ratcliffe 1993), when it was almost back to its prewar status. The situation was less straightforward with the sparrowhawk. For some time after the major restrictions on dieldrin, it continued to be used as a seed dressing for autumn-sown cereal crops in the main corn-growing areas of eastern England. It was only after a total ban on dieldrin as a seed dressing in 1975 that sparrowhawks began to return to the worst-affected area of eastern England.

The recolonization of this area only occurred after the levels of dieldrin in livers of sparrowhawks fell below 1 ppm (Newton and Wyllie 1992) (see Figure 9.9). As

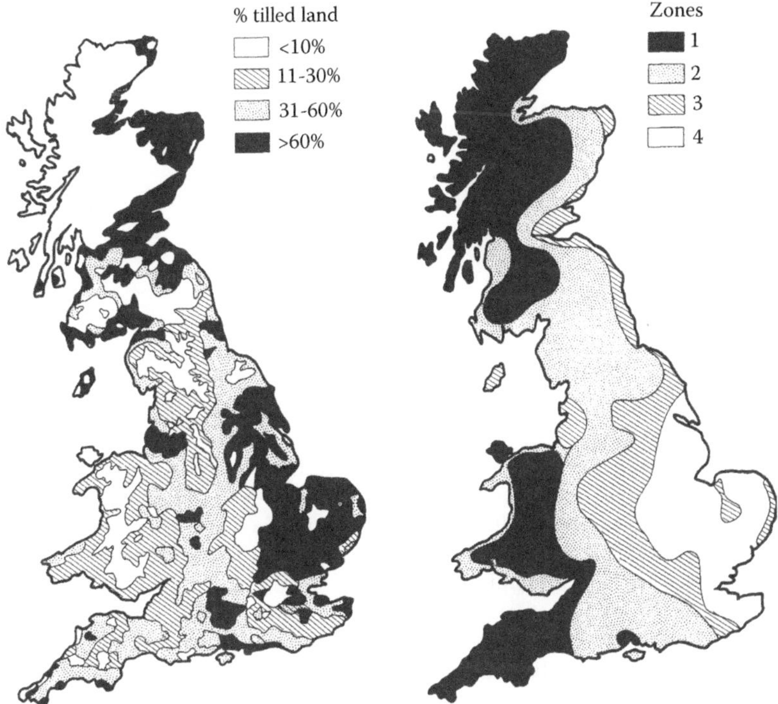

FIGURE 9.8 Changes in the status of sparrowhawks in relation to agricultural land use and organochlorine use. The agricultural map (left) indicates the proportion of tilled land, where almost all pesticide is used. The sparrowhawk map (right) shows the status of the species in different regions and time periods. Zone 1, sparrowhawks survived in greatest numbers through the height of the organochlorine era around 1960; population decline judged at less than 50% and recovery effectively complete before 1970. Zone 2, population decline more marked than in Zone 1, but recovered to more than 50% by 1980. Zone 4, population almost extinct around 1960, and little or no recovery evident by 1980. In general, population decline was most marked, and recovery latest, in areas with the greatest proportion of tilled land (based on agricultural statistics for 1966). Reproduced from Newton, I. and Haas, M. B. (1984). The return of the sparrowhawk. British Birds 77, 47-70; reproduced in Newton, I. (1986). *The Sparrowhawk.* Calton, Poyser. With permission from Academic Press.

we shall see, this is substantially below the concentration of dieldrin in the livers of birds of prey when they have been lethally poisoned. The recovery occurred despite p,p′-DDE remaining at a high enough level to cause significant eggshell thinning. This provided yet further evidence, if it were needed, that dieldrin rather than DDT was responsible for the decline of the sparrowhawk in Great Britain.

A critical part of the study of the effects of organochlorine insecticides upon birds has been the detection and determination of residues in tissues of individuals found dead in the field. The level of residue in tissues can give an indication of the toxic effect that a chemical has had. Many pesticides are not very stable; they are quickly broken down, especially by enzymes, once they enter living tissues. This makes it

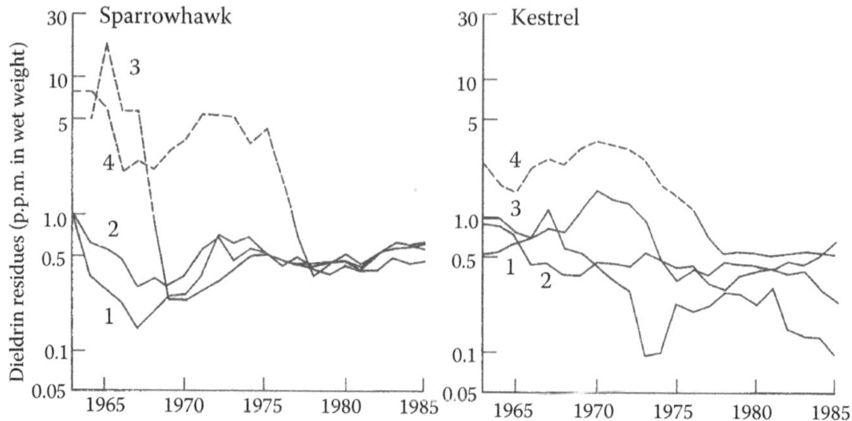

FIGURE 9.9 Number of sparrowhawk carcasses (—) receivedin a region of eastern Britain, together with concentrations of DDE (- - - - -) and dieldrin (– • – • –) found in their livers. Reproduced from Newton, I. and Wyllie, I. (1992). Recovery of a sparrowhawk population in relation to declining pesticide contamination. *Journal of Applied Ecology* 29, 476–484. With permission from Blackwell Science Ltd.

difficult to establish when they have toxic effects in the field. So, although persistence is an undesirable property of a pesticide from an ecological point of view—it has the advantage that evidence of toxicity is left behind for the analyst to find.

Of central interest at this time was what constitutes a lethal residue of dieldrin in birds of prey *under field conditions*. Field conditions are emphasized here because in the protected environment of the laboratory birds may be expected to survive higher exposures than under the harsher competitive conditions of the field—not least because neurotoxic compounds like organochlorines can impair their hunting skills. In the laboratory birds are presented with food and do not need to capture it for themselves. Residues of dieldrin + heptachlor epoxide found in dead sparrowhawks and kestrels in the field in Britain are presented in Figure 9.10. The heptachlor epoxide residues constitute only a few percent of the total residues; the reason for including them with dieldrin residues was that the two compounds are very similar, sharing the same mode of action, as explained earlier. The graph at the top left shows the dieldrin levels found in livers of dead kestrels in Great Britain during the period 1963–1975 when there was heavy use of cyclodienes. The plotting of the graph is on a logarithmic scale, and the peak centering on 20 ppm includes individuals observed to have died with symptoms of dieldrin poisoning. This may be compared with the results of a survey conducted in 1966–1967 in the province of Drente, the Netherlands, of residues in birds found dead after the sowing of seed dressed with dieldrin or heptachlor (Fuchs 1967). Statistical analysis of these data yields a geometric mean of 18.00 ppm for dieldrin plus heptachlor epoxide in the livers of buzzards (sixteen individuals) found dead. This is close to the value given earlier for kestrels evidently lethally poisoned by dieldrin in the field. There also appears to be a peak in this position in the graph for dieldrin residues in sparrowhawks found dead in eastern England in 1963–1986, although the sample size here

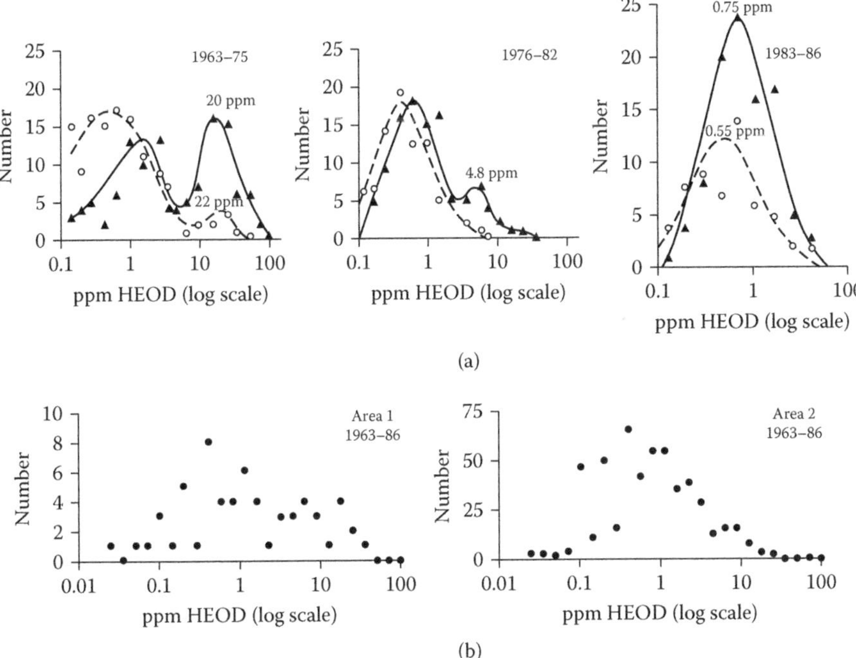

FIGURE 9.10 (a) Distribution of dieldrin (HEOD) residues in the livers of kestrels from two different areas of Britain. The HEOD residues are expressed as ppm wet weight. They are plotted on a log scale. The numbers of individuals with dieldrin residues falling within the ranges of concentrations are represented by 0.15 log units are given on the vertical axis. The concentrations plotted represent the midpoints of each log range. Area 1 (highest cyclodiene), D ▲; area 2 (lowest cyclodiene use), ○. (b) Distribution of HEOD residues in the livers of sparrowhawks. From Walker and Newton (1999).

is rather small and the peak is not as clearly defined as is the corresponding one for kestrels (Figure 9.10, bottom left-hand distribution diagram).

It is interesting to compare these figures with other data for liver residues in owls accidentally poisoned by dieldrin in the London zoo (Jones et al. 1978). Analyzing ten specimens of different species of owl, they reported a geometric mean of 28. This is higher than the figures obtained from raptors poisoned in the field. However, the lethal concentrations in captivity may be expected to be higher than those found in birds poisoned in the field. Birds in captivity affected by sublethal effects of the insecticide would be expected, for a time, to be able to consume food given to them even where they had lost the ability to hunt. In other words, birds in captivity would be able to survive higher doses of the chemical than would birds in the wild. Birds in the wild would lose the ability to hunt after receiving these lower doses associated with sublethal effects. In both cases, birds would eventually stop feeding, but this would happen at an earlier stage with the wild birds. In both cases, when feeding stopped, residues of dieldrin would be redistributed and liver levels would rise until the birds died. The captive birds would have been carrying a higher level and residue burden than the wild birds—and consequently would have been expected to carry higher liver residues at death than the wild birds.

To summarize, predatory birds poisoned in the field by cyclodienes typically contained a mean cyclodiene residue (dieldrin plus heptachlor epoxide) of about 20 ppm in their livers. A very large proportion of birds evidently poisoned contained residues of 9 ppm or more.

Returning to the critical study of Newton and Wyllie (1992), their field data suggested that sparrowhawk and kestrel populations were in a state of decline when the dieldrin liver levels exceeded 1 ppm. Recovery came when the levels fell below this figure. Sparrowhawks did not begin to recolonize former breeding areas of East Anglia (Figure 9.8, zone 4) until mean liver dieldrin levels fell below 1 ppm.

Looking again at the residues in kestrels (Figure 9.10, top), it can be seen that many birds contained apparently lethal concentrations of dieldrin (peak centering on 20 ppm) before the ban on seed dressing of 1975. However, in subsequent years (1976–1982) this peak disappeared, giving way to another centering on 4–5 ppm. There are reasons for suspecting that this latter peak represents individuals experiencing sublethal effects (Walker and Newton 1999; Walker 2004). Although the numbers of sparrowhawks are smaller, there appears to be a similar situation here in area 1, which represents zone 4, during the period 1963–1986. Over this period there appear to be two peaks in similar positions to those observed in the kestrel sample. Taking these observations together, it seems very probable that birds represented by peaks centering on 4–5 ppm would have experienced sublethal effects.

This observation may help to explain why the recovery of sparrowhawk and kestrels did not occur until mean dieldrin residue levels had fallen below 1 ppm. Individuals containing residues of 3 ppm and above are very likely to have experienced sublethal effects, and this could have meant starvation through impairment of hunting skills. This idea was put to the test using a population model for the sparrowhawk (Sibly et al. 2000). Based on the British field data two ideas were tested: (1) that the sparrowhawk had declined initially because of the lethal toxicity of dieldrin alone or (2) that this decline was due to both lethal and sublethal toxicity. It was assumed that individuals containing residues of 9 ppm or more died from lethal poisoning, but that individuals containing residues of 3–9 ppm died due to the consequences of sublethal poisoning. Using this model, it was found that estimated lethal toxicity alone was insufficient to account for the observed speed of the decline, whereas lethal plus sublethal toxicity predicted a decline that closely followed the one that actually occurred.

The assumption that birds containing 3–9 ppm died due to sublethal effects of dieldrin could then provide an explanation as to why sparrowhawks did not effectively recolonize the eastern area of England until their residues fell below 1 ppm. Above a mean level of 1 ppm there would still have been a significant percentage of individuals in the category 3–9 ppm, where sublethal effects are predicted. Studies upon humans as well as experimental animals have shown that dizziness, disorientation, drowsiness, hyperirritability, anorexia, and changes in electroencephalograph patterns are all features of sublethal poisoning by dieldrin and related compounds. It does not require a fertile imagination to appreciate the potential seriousness of such effects upon the hunting skills of avian predators such as sparrowhawks. In the best of circumstances they only have a low chance of catching their mobile avian prey.

Similar to these observations for the sparrowhawk—but on more limited evidence—it would appear that the peregrine may also have been affected in this way by dieldrin. During the same period, another bird-eating raptor, the merlin, experienced a decline believed to have been caused by dieldrin in northwest England (Newton et al. 1978).

DECLINE OF PREDATORY BIRDS RELATED TO EGGSHELL THINNING CAUSED BY P,P′-DDE

Although eggshell thinning associated with p,p′-DDE residues was reported in predators such as the peregrine, sparrowhawk, and kestrel in Great Britain, it was only related to limited and localized population declines. In North America, however, higher p,p′-DDE residues were reported in raptors and piscivorous birds than in the UK and some other parts of Europe—and greater degrees of eggshell thinning were found than in Western European studies. In general DDT residues were higher in North America than in Western Europe from the 1950s until bans came later in the twentieth century. The opposite was true with aldrin, dieldrin, and certain other cyclodienes. Residues of these were found at higher levels in the above species in Britain, the Netherlands, and Denmark than in the United States and Canada (Peakall 1993).

In North America during the late 1940s to late 1970s the decline of several species of birds of prey was associated with eggshell thinning caused by p,p′-DDE. Peregrine populations in several areas declined or were extirpated when thinning of 20–25% occurred. The decline of the bald eagle first reported in Florida in 1946 (Broley 1958) occurred in many areas of North America during 1946–1957. It was associated with eggshell thinning of 15–19% and diminished breeding success.

Eggshell thinning and consequent egg breakage and breeding failure were found in a number of piscivorous birds of North America, including the osprey, the double-crested cormorant, the brown pelican, and the gannet. The double-crested cormorant was studied on the Great Lakes of Canada and the United States. In 1972 and 1973 a detailed study of this species was conducted by the Canadian Wildlife Service (Weseloh et al. 1983). They found that eggshells were 24% thinner than prewar values, and that there was extensive nesting failure connected with a very high level of egg breakage. An even more dramatic example of egg thinning and breakage was in a brown pelican colony on Canapa Island off the coast of California in 1969. Here, eggshell thinning was of the order of 50% in many cases, and egg breakage was widespread. Reproductive failure was almost total—only four chicks were raised from some 750 nests.

Another study along similar lines was of a gannet colony on Bonaventura Island, Quebec, during the 1960s and early 1970s (Elliott et al. 1998). In 1969, poor breeding success was associated with severe eggshell thinning—and related to residues of p,p′-DDE of 19–30 ppm in the eggs. Subsequently, pollution of the St. Lawrence River by DDT was reduced. The p,p′-DDE levels in gannets fell, and by the late

1970s shells had become thicker, reproductive success increased, and the population recovered.

A common feature of all these studies is that egg breakage became serious only after eggshell thinning exceeded c. 15%. In peregrines this happened when thinning exceeded 20%. The severe declines of populations were found when p,p′-DDE residues became high enough to cause considerable shell thinning and extensive breakage. We have seen how this happened in a number of classic cases in North America. In Britain and Western Europe p,p′-DDE levels were seldom high enough to cause this problem. However, the cyclodiene insecticides, as we have seen, were another story.

OTHER EFFECTS OF ORGANOCHLORINES

Much of what is known about harmful side effects of organochlorine insecticides on natural populations was discovered retrospectively, after the damage had been done. Such was the case with many of the most serious effects upon predatory birds described above. With the benefit of hindsight, evidence has come to light suggesting other effects that were overlooked when they were actually happening. One such example was the disappearance of the otter from many British rivers over a similar period to the declines of sparrowhawks and peregrines. Chanin and Jefferies (1978) examined hunting records and related this decline in time and space to the introduction and usage of dieldrin. There is plenty of evidence that dieldrin reached substantial levels in fish and in piscivorous birds in these areas over this period. The otter is an aquatic predator feeding largely on fish. Of the very few otter carcasses that were found in the field at this time, some contained substantial levels of dieldrin, suggestive of sublethal or lethal toxic effects. With the bans on cyclodienes taking force during the 1960s and 1970s, dieldrin levels have fallen in British rivers and otters have returned to most of them.

Nearly all the examples given so far have been for the developed countries of the Western world. In developing countries, e.g., much of the African continent and the Indian subcontinent, the picture has been different. Here DDT and cyclodienes have been used, e.g., in vector control programs and in locust control after being banned in developed Western countries. In lands visited by famine and disease, the protection of the environment may seem an extravagant luxury. Koeman and Pennings (1970) reported on the use of dieldrin—and its highly toxic isomer, endrin—to control locusts and tsetse flies in Africa. With extensive spraying of these compounds large numbers of animals and birds were poisoned. DDT has continued to be used for controlling malarial mosquitoes in areas of India and Africa. See further discussion in Chapter 17.

During the early 1990s the International Atomic Energy Agency, supported by a grant from the Swedish government, investigated the misuse of pesticides in Africa (International Atomic Energy Agency/1997 report). Among other things, it was found that dieldrin continued to be used for termite control in parts of East Africa, and endrin had been used to aid catching fish in West Africa. DDE-related shell thinning was reported for certain predatory birds in East Africa.

RESISTANCE TO ORGANOCHLORINE INSECTICIDES

As described in Chapter 4, the widespread and continuing use of DDT for the control of many insect species led to the development of resistance. This has been studied in some detail in the housefly. Both target insensitivity and increased metabolism have been implicated. Some resistant strains show both types of mechanism. Target insensitivity has been classified as either kdr or super kdr. In both cases resistance is the consequence of the presence of a mutant form of the target site on the sodium channel of resistant insects. Metabolic resistance, on the other hand, is due to high DDT-ase activity in resistant insects. DDT-ase is the name given to an enzyme that can transform p,p′-DDT into its less toxic metabolite p,p′-DDE (Figure 9.1). *Toxicity* here refers to neurotoxicity to insects. It should not be confused with eggshell thinning in birds where the metabolite is more effective than the p,p′-DDE from which it is derived.

Insect resistance has also emerged toward cyclodiene insecticides. With dieldrin, resistance due to insensitivity of a mutant form of the target site has been reported for the housefly and some crop pests, e.g., cabbage root fly. The target site here, as described earlier, is the GABA receptor of the nervous system. The author is not aware of any examples of resistance to dieldrin that are due to enhanced metabolism. This is not altogether surprising because dieldrin is not very susceptible to metabolic attack (Figure 9.2). The high level of chlorination makes it difficult for oxidative enzymes to attack it, due to steric hindrance.

MORE ECOFRIENDLY CYCLODIENES

When the problems with dieldrin and certain other cyclodienes came to be recognized, moves were made to discover more ecofriendly compounds of this type. Some dieldrin analogues with lower levels of chlorination were synthesized and found to be metabolically degraded much more rapidly than dieldrin itself (Brooks 1969, 1974; Chipman and Walker 1979, Walker and E.I. Zorganic 1974, Brooks et al. 1970). The dieldrin analogues HCE and HEOM were examples of this, being susceptible to attack by both cytochrome P450-based oxygenases and the epoxide hydrolase of vertebrates (HCE is hexachloro 5,6 epoxyoctahydromethanonaphthalene and HEOM is hexachloro 6,7 epoxyoctahydromethanonaphthalene). However, with the very bad public image of organochlorine insecticides following the problems described earlier, these compounds were not developed commercially.

SUMMARY

The side effects of the neurotoxic organochlorine insecticides DDT, aldrin, and dieldrin attracted wide public attention during the 1960s. They increased public awareness of the potential hazards to the environment presented by persistent pesticides. One central issue was the strong persistence of certain of their residues in biota and in soils. Especially persistent were dieldrin and the DDT metabolite p,p′-DDE, which have long half-lives in animals and in soils, and can undergo biomagnification with progression along both terrestrial and aquatic food chains.

Although DDT and the cyclodiene compounds aldrin, dieldrin, and heptachlor are all neurotoxic, they differ in their modes of action. DDT acts upon a sodium channel that is located in nerve membranes, whereas the cyclodienes act upon a GABA receptor located on a chloride ion channel located in the vicinity of inhibitory synapses.

Biomagnification led to the appearance of significant concentrations of these residues in some predators occupying positions in the higher trophic levels of aquatic and terrestrial food chains. In Western Europe and North America there was evidence of population declines of predatory birds caused by the toxic effects of these residues.

The DDT metabolite p,p′-DDE was found to cause eggshell thinning and egg breakage in predatory birds such as the peregrine falcon, sparrowhawk, bald eagle, double-crested cormorant, gannet, and brown pelican. In North America, population declines of some these species were related to eggshell thinning caused by p,p′-DDE.

In Western Europe the declines of the sparrowhawk and peregrine falcon were attributed mainly to the toxicity of dieldrin. There was some evidence that sublethal effects of dieldrin may have contributed to these population declines. Although eggshell thinning caused by p,p′-DDE occurred in this region, the levels in birds were lower than in North America, and only small local declines of predators were related to this effect. There was some evidence that sublethal effects of dieldrin may have contributed to these declines.

Into the 1970s, bans and restrictions led to the decline of organochlorine residues in many areas and some population recoveries. Resistance to DDT and dieldrin developed in insects that had been exposed to them. The resistance mechanisms involved are described in the foregoing text.

During the 1970s some novel biodegradable cyclodienes were synthesized, but these were never developed commercially.

FURTHER READING

Broley, C.I. 1958. Plight of the American bald eagle. *Audubon Magazine* 60: 162–171. Early evidence of a decline in this species due to the effects of p,p′-DDE.

Brooks, G.T. 1974. *Chlorinated insecticides*. Boca Raton, FL: CRC Press. A very detailed and authoritative text on the chemistry and biochemistry of organochlorine insecticides.

Chipman, J.K., and Walker, C.H. 1979. The metabolism of dieldrin and two of its analogues: The relationship between rates of microsomal metabolism and rates of excretion of metabolites in the male rat. *Biochemical Pharmacology* 28: 1337–1445. Evidence that readily biodegradable analogues of dieldrin are rapidly eliminated by the rat.

Edwards, C.A. 1976. *Persistent pesticides in the environment*. 2nd ed. London: CRC Press. A valuable reference work on organochlorine insecticide residues in soils.

Elliott, J.E., Norstrom, R.J., and Keith, J.A. 1988. Organochlorines and eggshell thinning in northern gannets. *Environmental Pollution* 52: 81–102. A good example of a population decline due to DDE-induced eggshell thinning.

Hunt, E.G., and Bischoff, A.I. 1960. Inimical effects on wildlife of periodic DDD applications to Clear Lake. *California Fish and Game* 46: 91–106. A classic early paper describing the biomagnification of DDD in an aquatic food chain.

Moriarty, F. 1975. *Organochlorine insecticides: Persistent environmental pollutants*. London: Academic Press. A collection of in-depth reviews about this group of insecticides.

Newton, I. 1986. *The sparrowhawk*. Calton, UK: T & AD Poyser. An excellent monograph on the sparrowhawk that includes an account of the role of organochlorine insecticides in its decline.

Peakall, D.B. 1993. DDE-induced eggshell thinning: An environmental detective story. *Environmental Reviews* 1: 13–20. A valuable and insightful review containing much data about the peregrine falcon.

Ratcliffe, D.A. 1967. Decrease in eggshell weight in certain birds of prey. *Nature* 215: 208–210. A groundbreaking paper suggesting the involvement of DDT in eggshell thinning.

Ratcliffe, D.A. 1993. *The peregrine falcon*. Calton, UK: T & AD Poyser. A classical monograph with chapters about eggshell thinning and population decline related to organochlorine insecticides.

Walker, C.H. 2004. Organochlorine insecticides and raptors in Great Britain. In *Insect and bird interactions*, ed. H.F. Van Emden and M. Rothschild, 133–148. Andover, UK: Intercept Ltd. An account of the effects of these insecticides on raptors in Great Britain.

10 Organophosphorous and Carbamate Insecticides

INTRODUCTION

Following the environmental problems encountered with persistent organochlorine insecticides such as DDT, aldrin, dieldrin, and heptachlor, there was renewed interest in less persistent insecticides that might be suitable alternatives. Prominent among these were two groups of neurotoxic compounds that were readily biodegradable, the organophosphorous insecticides (OPs) and the insecticidal carbamates. Both types of insecticide act at the level of the nerve synapse as cholinesterase inhibitors.

The organophosphorous compounds were developed by both Germany and the Allies during and after World War II. Apart from their employment as insecticides, some were also considered as potential chemical warfare agents (nerve gases) (Fest and Schmidt 1982). Examples of the latter include soman, sarin, tabun, and diisopropyl phosphofluoridate (DFP) (Marrs et al. 2007). Fortunately these compounds were not employed as chemical weapons during World War II, although it is claimed that Saddam Hussein did use one of them as a chemical weapon against Kurdish villagers in Iraq. Recently, it appears that the nerve agent sarin has been used in the civil war in Syria.

The common mode of action of OPs and insecticidal carbamates will be explained before describing each group of insecticides separately.

MODE OF ACTION OF ANTICHOLINESTERASES

The mode of action of the active oxon form of an organophosphorous insecticide is illustrated in Figure 10.1. Here oxygen is attached to phosphorous by a double bond. The active forms of OPs are usually oxons. Often, as will be discussed later, commercial insecticides are thions, where sulfur, not oxygen, is attached to phosphorous by a double bond. Thions are more stable than oxons. Within living organisms the relatively stable thions are converted to the more reactive oxons, which are potent anticholinesterases.

Acetylcholinesterase is located in the postsynaptic membrane of cholinergic synapses (Figure 10.2). Synapses are junctions between nerves across which nerve impulses can be transmitted. At cholinergic ones the neurotransmitter is acetylcholine. When a signal arrives at the synapse, acetylcholine is released at the nerve ending to carry the message to the adjoining nerve. This molecule diffuses rapidly across the synaptic cleft and binds to an acetylcholine receptor on the other side (postsynaptic membrane). The act of binding causes a signal to be generated that is

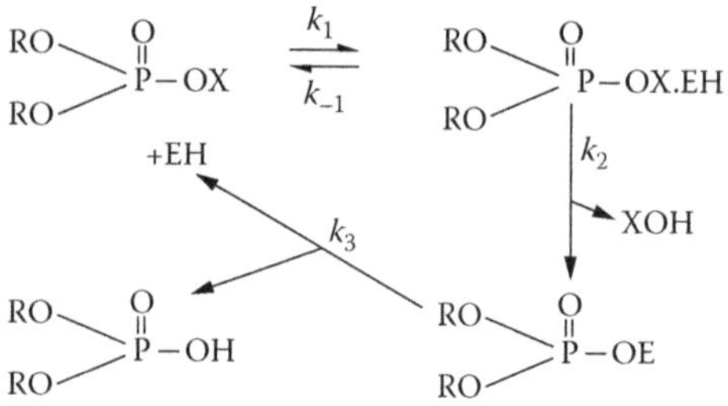

$$(RO)_2P(=O){-}OX + EH \underset{k_{-1}}{\overset{k_1}{\rightleftharpoons}} (RO)_2P(=O){-}OX.EH$$

$$(RO)_2P(=O){-}OX.EH \xrightarrow{k_2} (RO)_2P(=O){-}OE + XOH$$

$$(RO)_2P(=O){-}OE \xrightarrow{k_3} (RO)_2P(=O){-}OH + EH$$

FIGURE 10.1 Mode of action of organophosphorous oxons. R = alkyl group, E = enzyme.

rapidly carried along the second nerve. Neurotransmission across the synapse has been accomplished.

A vital characteristic of this system is that the transmission of the signal is quickly terminated. This is where acetylcholinesterase plays its part. It hydrolyzes the neurotransmitter acetylcholine, thereby terminating the signal. If, however, acetylcholinesterase is inhibited by an organophosporous compound or a carbamate, acetylcholine builds up in the synapse and signalling becomes continuous. If this situation persists, and signals continue to be sent, the acetylcholine receptor will run out of energy and be unable to continue operating. The synapse will be blocked, and it will be impossible to transmit signals from one side to the other. Synaptic block can lead to tetanus and respiratory failure.

In the diagram shown (Figure 10.1), the active oxon form interacts with acetylcholinesterase (EH). A phosphoryl residue binds to the active site of the enzyme with the displacement of a hydrogen ion. When in this state, the enzyme is unable to break down (hydrolyze) acetylcholine. Loss of this residue leads to reactivation of the enzyme, but this is usually a very slow process. If the phosphoryl residue undergoes modification while it is still bound to the enzyme, inhibition can become effectively irreversible. There are antidotes to anticholinesterase poisoning by OPs

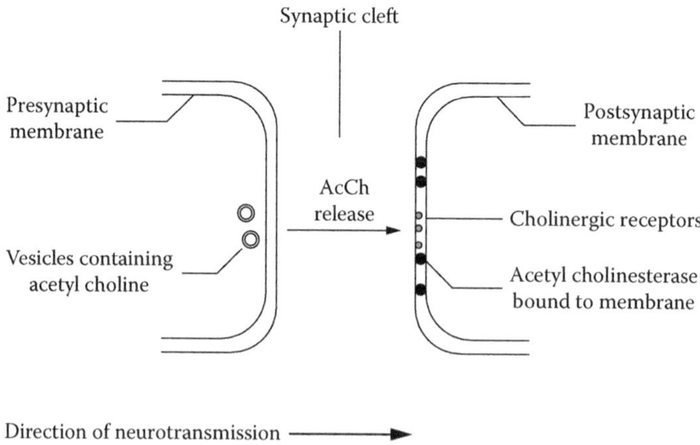

FIGURE 10.2 A cholinergic synapse.

such as pyridine aldoxime methiodide (PAM), which can reactivate inhibited cholinesterase. The natural product atropine is also used as an antidote, but acts by a different mechanism from PAM. It is an antagonist to acetylcholine at some (but not all) cholinergic receptors, and can thereby protect them from overstimulation by this neurotransmitter.

Insecticidal carbamates also bind to acetylcholinesterase at the same site as OPs. However, their binding is usually less strong and is readily reversible.

THE ORGANOPHOSPHOROUS INSECTICIDES

The organophosphorous insecticides (OPs) constitute a large group of compounds that vary considerably in their chemical and biological properties. Some of the compounds that will be described here are now restricted in use or completely banned in Western countries, although not necessarily elsewhere. The following account does not attempt to describe present-day usage of them worldwide, which would be a formidable task. It does, however, focus on evidence that exists of their harmful effects upon the natural environment when and where they have been used.

From a chemical point of view, structure, solubility, vapor pressure, and stability all vary considerably between individual compounds (Fest and Schmidt 1982; Hassall 1990). There is also great variation in their biological properties, e.g., biodegradability, toxicity, persistence, and systemic action in plants. The biological properties are related to the chemical ones (Ballantyne and Marrs 1992). Thus, compounds that are transported in the vascular system of plants (systemic insecticides) tend to have a significant degree of water solubility. This can apply to both original compounds and their metabolites. Systemic insecticides have been much used to control sap-feeding insects such as aphids and whitefly.

Compounds that are reasonably persistent after application in the field tend to have relatively low vapor pressures. Compounds showing marked selectivity (e.g., malathion, diazinon) often have structural features that determine the course of their metabolism. Compounds that cause delayed neurotoxicity have distinctive structural features. The last two points will be explained in more detail later. The basic structure of most OPs is illustrated in Figure 10.3, and the properties of some of them in Table 10.1.

As can be seen, four groups are attached to a phosphorous atom. R is often simply a methyl or ethyl group. Sometimes the link to phosphorous is through an oxygen bridge, for example, a CH_3O– (methoxy) or C_2H_5O– (ethoxy) group. The leaving group(X) is of relatively complex structure, and the fourth group is either O (oxon) or S (thion). Thions are more chemically stable than oxons, and many commercial OPs are thions. As we have seen, thions are converted to oxons by the action of oxidative enzymes within living organisms (cytochrome P450-based oxygenases), and they are strong inhibitors of acetylcholinesterase.

The leaving group X is so called because it breaks away from the rest of the molecule when an OP binds to acetylcholinesterase. The structure of X varies considerably between different insecticides. It strongly influences the character of each insecticide. A lipophilic leaving group makes for a lipophilic OP of low water solubility. Some leaving groups, such as that of malathion, include what is

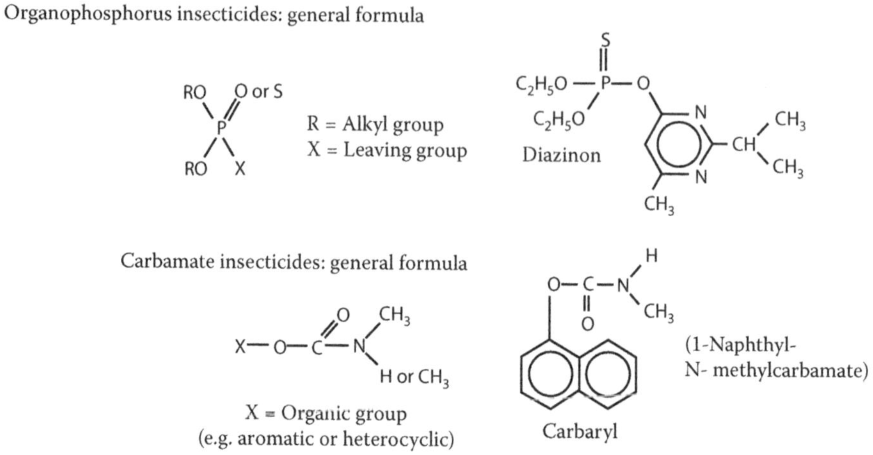

Organophosphorus insecticides: general formula

R = Alkyl group
X = Leaving group

Diazinon

Carbamate insecticides: general formula

X = Organic group
(e.g. aromatic or heterocyclic)

Carbaryl

(1-Naphthyl-
N- methylcarbamate)

FIGURE 10.3 Structures of organophosphorous and carbamate insecticide.

termed a selectophore. This is a subgroup that is available for the attack of detoxifying enzymes. Mammals readily detoxify malathion by esterase activity, whereas insects have only limited capability to carry out this reaction. Consequently, malathion is selectively toxic between insects and mammals. A similar situation exists with dimethoate.

SOME EXAMPLES OF ORGANOPHOSPHOROUS INSECTICIDES

Table 10.1 gives some examples of the OPs that illustrate the wide variation of properties and uses of this group of compounds.

TABLE 10.1
Some Organophosphorous Insecticides

Name	Water Solubility µg/ml	Vapor Pressure mmHg	Description
Diazinon	40	1.4×10^{-4}	Contact insecticide
Malathion	145	4.0×10^{-5}	Contact insecticide with some localized movement
Dimethoate	5000	8.5×10^{-6}	Systemic insecticide
Demeton-S-methyl	3300	3.6×10^{-4}	Systemic insecticide
Disyston	25	1.8×10^{-4}	Systemic insecticide Granular formulation
Chlorfenvinphos	145	3×10^{-6}	Contact insecticide Granular formulation. Also used as seed dressing

Source: Walker, C.H., *Organic Pollutants: An Ecotoxicological Perspective*, 2nd ed., Boca Raton, FL: Taylor & Francis, 2009.

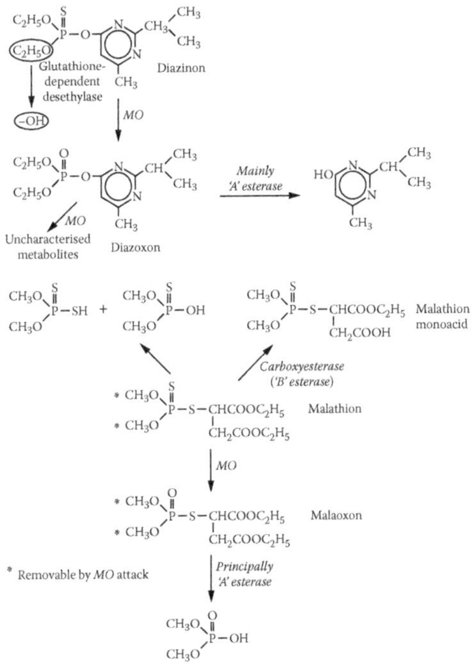

FIGURE 10.4 Metabolism of diazinon and malathion.

Diazinon is an OP that has been widely used. It has low water solubility and has been employed as a contact insecticide. The term *contact* indicates that it is not mobile after it is applied, but tends to express toxicity where it is originally deposited. By contrast, *systemic* insecticides readily move about in the vascular system of plants and can express toxicity in areas far removed from the original area of application (see Box 10.1). Its metabolism is summarized in Figure 10.4. Diazinon is a thion that is activated by conversion by cytochrome P450-mediated oxidation to diazoxon, its active oxon form. Three important detoxication pathways are shown. First, a glutathione-dependent enzyme can remove an ethyl (C_2H_5–) group from diazinon itself. Second, oxidative attack catalyzed by cytochrome P450 can produce several metabolites of diazopan. Third, an A esterase can break the ester bond that links the phosphoryl part of the diazoxon to its leaving group. All of these conversions cause loss of toxicity. The action of A esterase is of particular interest because it helps to explain the marked selective toxicity of diazinon between birds and mammals. Mammals have strong A esterase activity in the blood, whereas birds have only very weak activity. Diazoxon, together with the related oxons of pirimiphos-methyl and pirimiphos-ethyl, is highly susceptible to attack by this enzyme—and all three parent insecticides are much more toxic to birds than to mammals (Brealey et al. 1980; Mackness et al. 1987).

Malathion is another OP showing marked selectivity. It is very toxic to most insects, although there are resistant strains of some pests that are not susceptible to it. Like diazinon, it is essentially a contact poison, although it can move to a small

BOX 10.1 SYSTEMIC INSECTICIDES

Systemic insecticides are able to travel in the vascular system of plants, i.e., in the transpiration stream of the xylem or in the transportation stream of the phloem. Thus, insecticides taken up by the roots or across the cuticle of the leaf can reach other parts of the plant. This can be a very effective way of delivering insecticides to sap-feeding insects such as aphids or whitefly. Typically, the insecticides are formulated as emulsifiable concentrates or wettable powders that can be dispersed in water thus yielding sprays that are applied to aerial parts of crop plants. Sometimes they may be formulated as granules that slowly release their toxic constituents in the soil to be taken up by roots. Granules can also be applied to leaves where the insecticides can cross natural barriers to enter the vascular system. Granular formulations have the advantage that they are relatively safe to handle—because the insecticides that they contain are only slowly released.

Most systemic insecticides (or their active metabolites) have appreciable water solubility and are able to reach relatively high concentrations in the plant vascular system. Examples given here include the OPs dimethoate and demeton-S-methyl. Some other systemic OPs, e.g., disyston and phorate (Table 10.1), have low water solubility but are metabolized to form relatively stable and water-soluble metabolites within plants. These metabolites are very toxic to sap-feeding insects.

A few carbamate insecticides are also systemic, e.g., carbofuran and aldicarb. Typically, these are formulated as granules.

Some fungicides are also systemic, e.g., carbendazim, prochloraz, ketoconazole, imazalil, and prochloraz. Many of these compounds act as ergosterol biosynthesis inhibitors (EBIs).

extent in the lipid membranes of plant leaves. Mammalian toxicity is low, and it has been used in human medicine to control insect parasites. The selective toxicity is again due to differential metabolic detoxication. Malathion, like diazinon, is a thion that is oxidized within living organisms to yield its toxic oxon, malaoxon. In the leaving group of malathion there are two carboxy ester bonds that can be hydrolyzed by certain esterases. The cleavage of one or both of these groups leads to detoxication. Mammals are not very susceptible to malathion because they possess carboxylesterases that can rapidly detoxify malathion. Insects are deficient in this capability and are much more susceptible. Both malathion and diazinon illustrate how it is possible, by intelligent molecular design, to produce new selective insecticides that are more ecofriendly than their predecessors.

Demeton-S-methyl and dimethoate are systemic insecticides, and this property is related to their relatively high water solubility in comparison to contact OPs such as diazinon and malathion. Disyston appears to contradict this principle because it is a systemic insecticide of low water solubility. However, the devil, as usual, is in the details. Disyston itself does not reach high concentrations in the vascular system of plants, but it is converted into toxic metabolites (sulfones and sulfoxides) inside plants. These are relatively water soluble and move readily in the vascular system to give control of sucking insects such as aphids over a significant period. Disyston has high mammalian toxicity and is formulated as granules that slowly release the insecticide in the field. Chlorfenvinphos is a relatively stable OP with a rather low vapor pressure. Consequently, it is more persistent than most OPs. When restrictions were placed on organochlorine insecticides during the 1960s and 1970s, it was employed as an alternative to dieldrin as a seed dressing.

TOXIC EFFECTS OF OPS

Most of the toxic effects of OPs on animals are due to inhibition of acetylcholinesterase of the nervous system. In vertebrate animals and birds, effects upon this enzyme in the brain are of particular interest and importance. Effects upon acetylcholinesterase of the peripheral nervous system are also important. The dose-dependent progression of the effects of OPs on animals are illustrated in Figure 10.5.

With increasing doses of an OP a sequence of toxic effects are seen. At low doses no effects are evident, but with higher doses mild sublethal effects begin to appear, and with further increases in dose these are replaced by more severe sublethal effects. Up to this point effects are reversible and will disappear if dosing is discontinued. At higher doses still, effects are severe and irreversible and lead to death. This sequence follows a progressive increase in the inhibition of brain cholinesterase. A typical sequence in birds dosed with OPs is the appearance of mild effects at 40–50% inhibition of brain cholinesterase, more serious effects at inhibition levels above this, and finally, severe, irreversible effects and death above 70% inhibition (Grue et al. 1991). There can be considerable variation between individual subjects in the relationship between the degree of brain cholinesterase inhibition and the effect. The inhibition

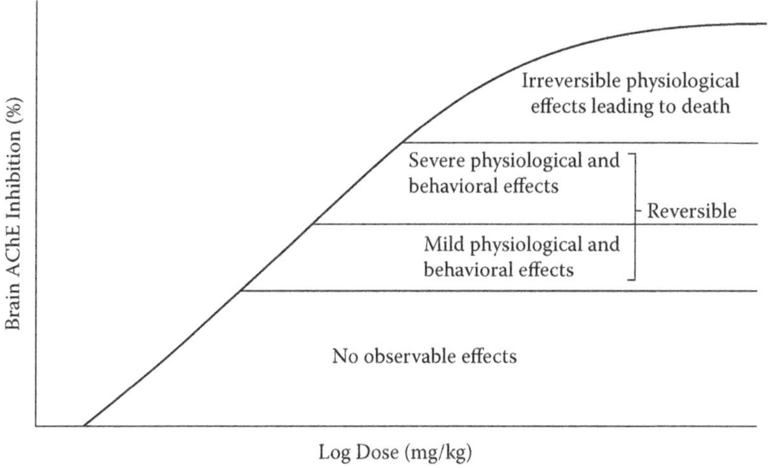

FIGURE 10.5 Stages in OP intoxication.

of acetylcholine esterase provides the basis for a valuable *mechanistic biomarker assay* for measuring the effects of OPs upon animals and birds in the natural environment. Examples will be given in the following text.

As with nerve poisons (neurotoxins), more generally there is concern about possible sublethal effects of OPs upon behavior. That such effects can arise has become evident during the course of toxicity tests with vertebrates. Studies with fish have shown that behavioral changes can occur at quite low levels of cholinesterase inhibition. For example, Beauvais et al. (2000) observed behavioral changes in rainbow trout when exposed to sublethal levels of diazinon and malathion in ambient water. Some of these effects were detectable down to only 20% inhibition of brain acetylcholinesterase. Effects included changes in speed and duration of swimming. This is a significant finding because although behavioral effects may escape notice, they can critically reduce chances of survival—for example, by significantly reducing the chances of a fish avoiding predation. Behavioral effects can be important if they affect the hunting skill of predators or, for example, the ability of honeybees to communicate with one another through their "dance," their success in finding food, or in mating.

Cholinesterases are not the only target of OPs. A few OPs can inhibit another esterase of the vertebrate nervous system that is termed neuropathy target esterase (NTE) (Johnson 1992). This esterase came to light during the investigation of delayed neuropathy in humans (ginger-jake paralysis) caused by triorthocresolphosphate, a compound present in an illicit liquor consumed in the United States during the period of prohibition. Subsequent studies showed that the symptoms did not appear until some two–three weeks after exposure. Later investigation showed that a few OPs, including DFP, mipafox, leptophos, and methamidofox, could cause delayed neuropathy by the same mechanism. Apart from humans, cats, pigs, sheep, and chickens are also known to be susceptible to this toxic effect, but it is not known whether the named OPs have had such effects in the natural environment. Insecticides known to cause delayed neuropathy were withdrawn from the market when the problem was recognized.

A biochemical study of preparations of rat brain showed that diazoxon bound to two sites other than acetylcholinesterase. Moreover, it showed a stronger affinity for these sites than for acetylcholinesterase itself. The OP dichlorvos also showed a strong affinity for one of these sites (Richards et al. 1999). This work has raised the question whether interaction with either of these sites can lead to toxic effects on rats or other vertebrates.

EFFECTS OF OPS IN THE NATURAL ENVIRONMENT

OPs were once very widely used for a variety of purposes, but with the introduction of much stricter controls, especially in developed countries, they are not so much used today. Common uses have been as sprays, granules, dusts, and seed dressings to control crop pests. In veterinary practice they have been used to control ectoparasites of farm animals, as in sheep dipping. Aerial spraying has been carried out in North America to control pests in orchards and forests, and in Africa against locusts and the avian pest, *Quelea*. The list is a long one.

Some OPs have presented a serious hazard to birds and other vertebrates in the field because of their acute lethal toxicity. However, these compounds are not as persistent in either the biotic or abiotic environment as DDT, DDE, or dieldrin. With the possible exception of a few compounds causing delayed neuropathy, their harmful effects appear to have been short-lived in comparison with those caused by the persistent organochlorine insecticides.

In California, OP sprays have been used to control pests in almond orchards, which has led to the lethal poisoning of red-tailed hawks (*Buteo jamaicensis*) (see Hooper et al. 1989). In New Brunswick, Canada, there was an extensive forest aerial spraying program over many years during the late twentieth century to control spruce bud worm. A number of insecticides were employed over different periods (Peakall and Bart 1983), including the OPs phosphamidon and fenitrothion. Cholinesterase inhibition was measured in birds found dead after spraying, and it was estimated that nearly 3 million birds were poisoned in 1975 alone, largely due to the effect of phosphamidon. The effect of fenitrothion was less dramatic. A very large area was sprayed during this program.

Another large field kill of birds was observed in central Scotland during 1971–1972 (Hamilton et al. 1976). Hundreds of migrating geese (*Anser* species) were poisoned by the OP seed dressing carbophenothion. One of these species was the greylag goose. It was estimated that about two-thirds of the entire British wintering population of this species visited this area in the winter during this period. This highlights a problem with field incidents of this kind. If a large proportion of an entire population migrates through an area where there is a high risk of poisoning, there is a serious threat to the species as a whole.

RESISTANCE TO OPS

There are many examples of resistance to OPs developed by insects. The most common mechanisms are insensitivity of target site (i.e., acetylcholine esterase) and enhanced detoxication (see Chapter 5). Considering target site insensitivity first, many examples are known of resistant strains of pest species that possess aberrant (mutant) forms of acetylcholinesterase. Such mutant forms are rare in the general susceptible population. Sequence analysis of the gene coding for this enzyme in OP-resistant strains of the housefly and the fruit fly have identified six mutations that are associated with OP resistance. All of them code for amino acids that are situated near the active center of the enzyme (Salgado 1999; Devonshire et al. 1998). Sometimes a change in an amino acid located near the active center can obstruct the movement of the active form of an OP, preventing it from reaching the active center of the enzyme, without restricting access of the substrate acetylcholine. Thus, a mutant form of acetylcholinesterase can retain its physiological function of breaking down acetylcholine while being insensitive to an OP.

Enhanced metabolic detoxication is another important mechanism of resistance to OPs. Mention has already been made of resistance shown by clones of aphids due to high levels of a carboxylesterase (Chapter 5). Different clones of these aphids have different numbers of copies of the gene encoding for this enzyme. Clones that have large numbers of copies of this esterase gene can bind and break down active forms of OPs, thus conferring resistance. Other types of enzyme that have also been implicated in the resistance of insects to OPs include certain forms of cytochrome P450 monooxygenase and glutathione-S- transferase.

CARBAMATE INSECTICIDES

Like the pyrethroids and the neonicotinoids, the carbamate insecticides have been modeled on a natural product, in this case physostigmine. Physostigmine is a constituent of the Calabar bean, which has been employed as a "truth drug" by African tribes. In witchcraft trials, the accused was made to drink a suspension in water of powdered beans. In theory, if the accused confidently drank a large draft, he would promptly be sick, thus eliminating most of the poison from the body. A guilty party would drink only small amounts, which were duly ingested, thus effectively absorbing the poison, often with fatal consequences.

Physostigmine (Figure 10.6) is a carbamate that strongly inhibits acetylcholinesterase (i.e., it is an anticholinesterase), which has been employed in human medicine. It bears a structural resemblance to acetylcholine. The insecticidal carbamates (Figure 10.6) have been modeled on its structure (Kuhr and Dorough). Carbaryl has been widely used as a contact insecticide (Table 10.2). It is not very persistent, being quite volatile and being broken down quite rapidly by the cytochrome P450-based monooxygenases. It has relatively low mammalian toxicity and was one of the safer alternative insecticides brought in to replace the persistent organochlorine insecticides for certain purposes. On account of its susceptibility to oxidative detoxication, it can be strongly synergized by piperonyl butoxide and related compounds (Box 2.1).

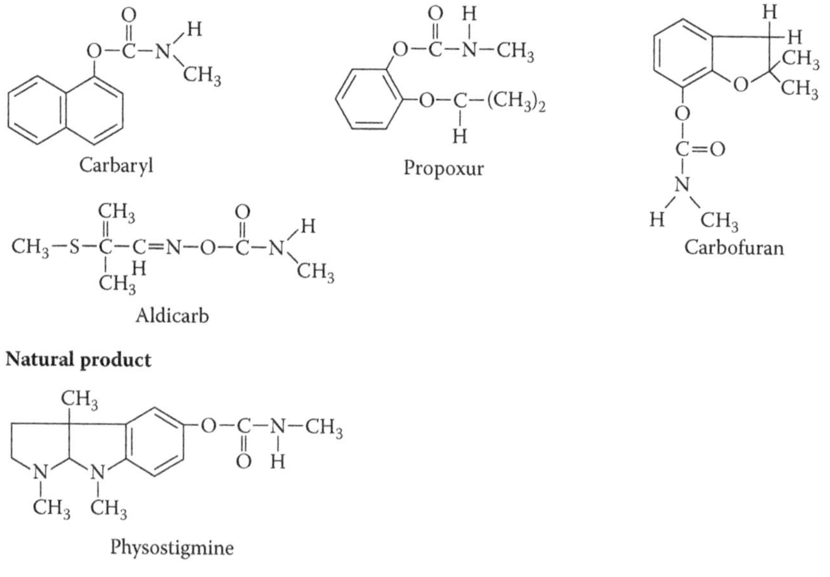

Carbaryl

Propoxur

Carbofuran

Aldicarb

Natural product

Physostigmine

FIGURE 10.6 Some insecticidal carbamates.

TABLE 10.2
Properties of Some Carbamates

Name	Water Solubility mg/L	Vapor Pressure mmHg	Toxicity Acute Oral LD_{50} (rat)
Carbaryl	40	3×10^{-3}	200–970
Aldicarb	6000	1×10^{-4}	0.1–7.7
Carbofuran	700	1×10^{-5}	6–14

Source: Walker (2009). *Organic Pollutants: an Ecotoxicological Perspective*, 2nd Ed., Boca Raton: Taylor and Francis.

Two other carbamates, aldicarb and carbofuran (Figure 10.3), are systemic and have high mammalian toxicity. In consequence, they have been formulated as granules to make them safer to handle. Like the systemic OPs dimethoate and demeton-S-methyl, they have considerable water solubility in contrast to carbaryl, which is not systemic.

Carbamates in general tend to be more toxic to birds than they are to rodents (Table 10.2). It has been suggested that this may be related to more rapid detoxication by monooxygenases in rodents than in birds (Walker 1983).

TOXICITY OF CARBAMATES IN THE FIELD

There have been field reports of heavy mortalities of birds and mammals on agricultural land caused by granular or pelleted formulations of highly toxic carbamates. In one case, c. 100 black-headed gulls were poisoned when a granular formulation of aldicarb was used (Hardy 1990). This occurred under wet conditions. Evidently, the birds were poisoned both by consuming contaminated earthworms and insecticidal granules. In another incident, the highly toxic carbamate methiocarb was applied as a pelleted formulation to control slugs (Greig-Smith et al. 1992). Wood mice were lethally poisoned, and in the short term, there was a marked population decline.

It has also been shown, in heavy soils, that significant quantities of carbofuran can be eluted from granules, then through the soil profile and into neighboring water courses. This was found when clay soils developed fissures during drought, and subsequent heavy rain took dissolved carbofuran through in the drainage water. Levels in drainage water were high enough to lethally poison deployed freshwater shrimp (Matthiessen et al. 1995).

LONGER-TERM EFFECTS ON SOIL MICROORGANISMS

The long-term use of carbofuran and other carbamates has been associated with changes in the metabolic capacity of soil microorganisms. In what have been described as problem soils, carbofuran granules were found to lose their effectiveness if regularly used (Suett 1986). This may have been due to the increase in numbers of certain species or strains capable of using carbofuran as an energy source. It is also possible that with a rise in concentration of the insecticide in soils, there is an induction of one or more enzymes present in microorganisms that can degrade it. A similar phenomenon has been identified with some herbicides, such as methoxychlorophenoxycetic acid (MCPA) and 2,4-D.

FURTHER READING

Ballantyne, B., and Marrs, T.C. 1992. *Clinical and experimental toxicology of organophosphates and carbamates*. Oxford: Butterworth/Heinemann. A valuable reference work on these two groups of insecticides.

Grue, C.E., Hart, A.D.M., and Mineau, P. 1991. Biological consequences of depressed cholinesterase activity in wildlife. In *Cholinesterase inhibiting insecticides—Their impact on wildlife and the environment*, ed. P. Mineau, 151–210. Amsterdam: Elsevier. A useful review on the effects of OPs on wildlife.

Kuhr, R.J., and Dorough, H.W. 1977. *Carbamate insecticides*. Boca Raton, FL: CRC Press. A valuable reference work on carbamate insecticides.

Matthiessen, P., Sheahan, D., Harrison, R., et al. 1995. Use of a *Gammarus pulex* assay to measure the effects of transient carbofuran run off from farmland. *Ecootoxicology and Environmental Safety* 30: 111–119.

11 Organometallic Compounds

INTRODUCTION

Thus far, most of the pollutants considered have been either organic or inorganic molecules. In this chapter we shall turn to some other molecules that are both organic and inorganic. Some metals, including mercury, lead, and tin, can form strong bonds with organic groups such as methyl, ethyl, or phenyl. The same is true of the metalloids arsenic and antimony. The properties of a metal or a metalloid can be radically changed with the attachment of these organic groups. In particular, groups such as ethyl, methyl, or phenyl have little charge and tend to reduce polarity. A consequence of this is that the organometallic compounds tend to have greater lipophilicity (fat solubility) than the inorganic compounds/ions from which they are derived. As a consequence, organometallic compounds can behave very differently and have very different toxicological properties from the inorganic compounds/ions from which they are derived.

This characteristic is illustrated by the mammalian toxicology of organolead and organomercury compounds that, because of their lipophilicity, readily cross the blood brain barrier and cause brain damage. Their inorganic precursors have less tendency to do this, but do cause damage elsewhere in the body, e.g., to tissues such as the kidney and the heart, which are not protected by the blood brain barrier.

As explained earlier (introduction to Chapter 1), organomercury and organoarsenic compounds are both naturally occurring and man-made. So when residues of them are detected in the environment, it is difficult to establish to what extent they are of natural origin and to what extent anthropogenic. Organometallic compounds have been produced commercially for various purposes. Organotin, organomercury, and organoarsenic compounds have all been used as pesticides. Organolead compounds have been used as antiknocks added to petrol.

TRIBUTYL TIN COMPOUNDS

Tributyl tin (TBT) compounds have been used as biocides and molluscicides. In the following account tributyl tin oxide, one of the best known, will be used as an example. Its structure is as follows:

$$(C_4H_9)_3\,Sn - O - Sn\,(C_4H_9)_3$$

It is a colorless liquid of low water solubility. It was once widely used as a component of antifouling paints that were applied to boats of many kinds, ranging from small leisure craft to large oceangoing vessels. Release of TBT from antifouling paints has provided a small but highly significant source of aquatic pollution.

An early indication of the problems associated with TBT came from stretches of the French coast, most notably the Bay of Arcachon, during the late 1970s (Alzieu 2000). Oyster populations began to show evidence of harmful effects. Poor shell growth, shell malformations, and very poor spatfalls were noticed. In time, these effects were attributed largely or entirely to TBT. Affected areas had relatively large numbers of small boats—and significant levels of pollution by TBT compounds. There followed a ban on the use of TBT compounds on small boats and a consequent fall in the level of this pollutant in coastal waters. Oyster populations recovered following the ban. Since the discovery of these effects in France, harmful effects of TBT compounds have been reported from other parts of the world, including the United States, Japan, and other Western European countries.

Later on across the channel, along the coast off southern England, a decline of the dog whelk population was noticed, and this too was associated with relatively high levels of TBT in seawater. These molluscs disappeared completely from certain shallow waters where there were large numbers of small boats, e.g., in harbors, marinas, and estuaries (Matthiessen and Gibbs 1998). Once again, it was found that TBT was having a harmful effect upon the molluscs. Dog whelks were bioconcentrating TBT from seawater. Females were developing male characteristics and, consequently, becoming infertile. More specifically, the development of a penis by females was causing blockage of the oviduct, making fertilization impossible. Following these discoveries, a ban was placed upon the use of paints containing TBT on small boats. The condition became known as imposex, and provided the basis for a biomarker assay that detects the effects of TBT on the dog whelk in field and laboratory studies.

Further evidence has implicated TBT in harmful effects caused to other species of gastropod mollusc. The sting whelk and the common whelk were both found to develop imposex in a similar way to the dog whelk when exposed to the chemical. The periwinkle showed gross malformation of the oviduct, a condition termed intersex, when exposed to TBT. There is evidence suggesting that these effects are the consequence of increased levels of the hormone testosterone in affected species, and that this is due to inhibition of the enzyme aromatase by the molluscicide. (Aromatase is a monooxygenase enzyme that contains cytochrome P450 and metabolizes testosterone.) For further discussion see Matthiessen and Gibbs (1998).

This type of endocrine disruption has been reported in more than 100 species of gastropod mollusc worldwide, a phenomenon that will be discussed further in Chapter 15.

ORGANOMERCURY COMPOUNDS

Apart from their production commercially, methylmercury compounds are synthesized from inorganic mercury by microorganisms in the natural environment. This process can be carried out by certain bacteria existing in sediments as part of a cyclic process (see Figure 11.1). Elemental mercury (Hg) in sediments is converted chemically into the mercuric ion, Hg^{2+}. This ion is then methylated by the action of bacteria to produce the methylmercury cation, CH_3Hg^+. Further bacterial methylation

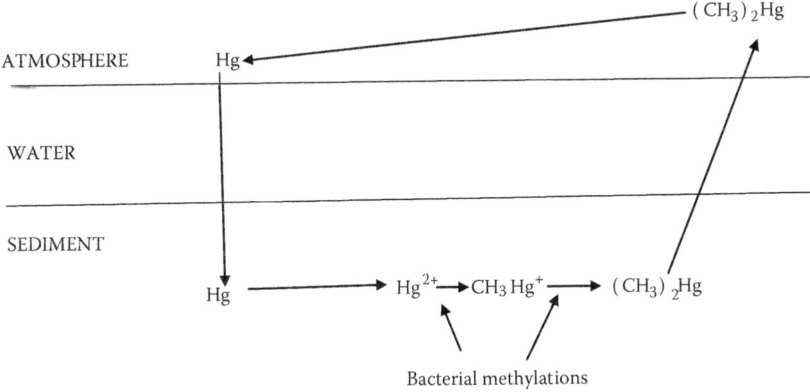

FIGURE 11.1 The cycling of mercury.

yields dimethylmercury, $(CH_3)_2Hg$, which is volatile and moves into air. Ultraviolet light breaks down dimethylmercury to release elemental mercury. Then elemental mercury can dissolve in water to complete the cycle. It is noteworthy that significant levels of methylmercury were present in the environment before it was synthesized by chemical industry. Organomercury compounds were not produced commercially until early in the twentieth century, but methylmercury has been found in the tissues of tinned tuna fish predating this period.

Organomercury compounds were widely used as agricultural fungicides in the middle of the twentieth century. They were also used as antifungal agents in the wood and paper industries. As with certain organochlorine insecticides (e.g., dieldrin), they were commonly used as seed dressings. The general formula for organomercury compounds is R-Hg-X, where R is a tightly bound organic group such as methyl (CH_3–) or phenyl (C_6H_5–), and the X group is usually an inorganic ion such as chloride (Cl^-). The solubility of organomercury compounds is mainly determined by X groups. Nitrates and sulfates tend to be more polar and water soluble than chlorides. The R-Hg bond is stable chemically and is not split by water or weak acids or bases. It is, however, more readily broken by enzymic attack, thus releasing inorganic mercury.

In contrast to methylmercury compounds, other organomercurials that have been produced commercially are not known to be biosynthesized from inorganic mercury in the natural environment. Prominent among these are phenylmercury and ethylmercury compounds. In general, these organomercury compounds are known to be highly toxic to mammals and birds. Acute oral LD_{50} values fall in the range of 5–70 mg/kg for mammals. The methyl compounds, however, are more persistent in the environment than the other types and, as we shall see, have been regarded as the most hazardous to wildlife. Biodegradation of organomercury compounds involves the enzymic removal of the organic group from inorganic mercury, and this process occurs more slowly in methylmercury compounds than with phenyl or ethyl ones. The ensuing account will deal mainly with methylmercury, with occasional reference to other forms of organomercury.

BOX 11.1 THE MINAMATA ENVIRONMENTAL DISASTER

The neurotoxicity of methylmercury was dramatically illustrated in the region of Minamata Bay, Japan, in the late 1950s. The release of both inorganic and methylmercury from a neighboring chemical factory led to the appearance of high levels of the latter in the marine ecosystem. Methylmercury underwent biomagnification in the marine food chain. Levels in fish and other seafood were high enough to cause lethal intoxication of some local people for whom these were the principal protein source. Some people died as a consequence of brain damage caused by methylmercury. Lethal brain levels were in excess of 50 ppm. Biomagnification of methylmercury in the marine food chain was an important factor in this disaster.

Like organotin compounds, organomercury compounds are markedly liposoluble and can cross the blood brain barrier. In the brain they combine with sulfydryl groups of proteins, thereby causing brain damage. The tragic consequences of this became manifest in an environmental disaster in the region of Minamata Bay, Japan, when many people were poisoned by methylmercury (see Box 11.1).

THE BIOMAGNIFICATION OF METHYLMERCURY

The biomagnification of methylmercury in food chains can be likened to that of the persistent organochlorine insecticides (Chapter 10), although the reported biomagnification factors have not been so large as those found for p,p'-DDE or dieldrin. As with the persistent organochlorine insecticide residues, methylmercury compounds combine lipophilicity with slow biodegradation. In a study with captive goshawks (*Accipiter gentilis*) (Borg et al. 1970), chickens were fed grain dressed with methylmercury, and tissue from these chickens was fed to the goshawks. The results are summarized in Table 11.1. Please note that the term *bioaccumulation factor* is used here because the transfer of methylmercury between species in this study was due

TABLE 11.1
Bioaccumulation of Methylmercury in a Laboratory Study

Material/Species	ppm Methylmercury	Duration of Feeding (days)	Approximate Bioaccumulation Factor
Dressed grain	8	N/A	N/A
Chicken tissue fed to goshawks	10–13	c. 40 days	×1.5
Muscle of goshawks fed chicken tissue	40–50	30–47 days	×4

Source: Data from Borg, K., et al., *Environmental Pollution* 1: 91–104, 1970.

only to ingestion of the chemical with food. In the natural environment the situation is more complicated and organisms can acquire residues by other means. In the aquatic environment chemicals can be taken up directly from ambient water as well as from food. In the terrestrial environment animals can consume contaminated water, soil invertebrates can absorb chemicals across the integument, etc. For this reason, the term *biomagnification* has been reserved for increases in residue concentrations with progression along food chains. Both bioconcentration and bioaccumulation can contribute to the biomagnification of methylmercury.

Summarizing the data in Table 11.1, chickens feeding on dressed grain bioaccumulated the methylmercury by a factor of c. 1.5. The goshawks bioaccumulated the methylmercury in their dietary chicken by about fourfold. Thus, if we look at the overall bioaccumulation factor, comparing levels in goshawks with those in the dressed grain, this is approximately sixfold.

POISONING OF PREDATORY BIRDS IN THE FIELD

During the 1960s there were reports of predatory birds being lethally poisoned by methylmercury seed dressings in the field (Environmental Health Criteria 86 published by the World Health Organization (WHO) in 1989). Many of these came from Northern Europe, including Sweden and the Netherlands, where methylmercury was widely used as a seed dressing. This did not appear to be a serious problem in the United Kingdom at the time because phenylmercury compounds rather than methylmercury compounds were used as fungicidal seed dressings. This evidence of secondary poisoning led to the banning of methylmercury seed dressings in Sweden and some other countries. The use of the less hazardous phenylmercury fungicides continued in Western countries for some time after these bans on methylmercury.

A retrospective study of levels of mercury in the feathers of Swedish birds was conducted by Berg et al. (1966). Samples dating from the 1840s to the 1950s contained relatively constant levels of the element. However, there was a sharp rise in concentration of 10- to 20-fold that coincided with the introduction of alkylmercury seed dressings (methyl and ethyl groups are examples of alkyl groups). The authors concluded that this rise was a consequence of the use of alkylmercury seed dressings.

METHYLMERCURY LEVELS IN PISCIVOROUS VERTEBRATES IN NORTH AMERICA

A number of studies conducted since the introduction of various bans on the marketing of organomercury compounds in North America have shown that levels of methylmercury high enough to cause physiological/biochemical effects continue to exist in aquatic ecosystems (Wolfe et al. 1998, 2007). Here, as in studies of other areas, such as the Mediterranean Sea, high levels of mercury have been found in the brain. Usually >80% of this has been methylmercury. It has been suggested that much of the methylmercury found in free-living vertebrates has been biosynthesized from inorganic mercury by microorganisms, rather than originating from synthesis by man.

A piscivorous bird containing relatively high levels of methylmercury that has been extensively studied in lakes of the United States is the common loon (*Gavia immer*) (Evers et al. 1998). In one study mercury levels were measured in birds and related to reproductive parameters. There was some evidence for reduced reproductive success when female loons contained blood concentrations of mercury above 3 ppm. The relationship between blood concentrations and behavior was also investigated. It was found that high-energy behaviors were reduced with increasing concentrations of mercury in the birds. These high-energy behaviors included foraging for chicks, foraging for self, swimming, flying, preening, and agonistic behaviors. In an associated laboratory study it was found that there were adverse effects upon loon chicks when they were dosed with food containing methylmercury levels comparable to the highest level found in the field study.

In Canada substantial levels of mercury (largely methylmercury) have been found in wild mink (<5.7 ppm in brain) (Basu et al. 2005). The mink were trapped in Yukon territory, Ontario, and Nova Scotia. The levels of mercury in the brain were found to be related to the abundance of certain receptors of the brain. These were muscarinic receptors and dopaminergic receptors, which respond to neurotransmitters. If this is a causal relationship, it suggests that organomercury may affect brain function, which might relate to the behavioral disturbances found in loons.

In summary, significant methylmercury residues continue to be reported in vertebrates from the highest levels of aquatic food chains in North America, and there is some evidence that they may be having adverse effects.

ORGANOLEAD COMPOUNDS

Lead tetraalkyl compounds, of which lead tetramethyls and lead tetraethyls are the best known, were once widely used as antiknock compounds in petrol. In other words, they were used to control semiexplosive burning in internal combustion engines. During the latter years of the twentieth century there was a substantial reduction in the use of leaded petrol because of the perceived health risk to humans (see Chapter 8 in this book; Bryce-Smith 1971).

Lead tetraalkyls are fat-soluble liquids of low water solubility. Like organomercury compounds, they can be degraded by enzymic attack. The first step in their degradation leads to the release of a trialkyl lead anion, as in the following example for tetraethyl lead:

$$Pb(C_2H_5)_4 \rightarrow Pb(C_2H_5)^{3+}$$

The transformation is carried out by an oxidative system based upon cytochrome P450.

There has been some evidence of harmful effects of lead tetraalkyl compounds causing damage in the natural environment. In 1979 a major poisoning incident occurred in the Mersey estuary, United Kingdom (Bull et al. 1983), caused by effluent containing tetraalkyl lead from a petrochemical works. Over 2,000 birds were found dead, including different species of gulls, waders, and ducks. The majority of them were dunlin (*Calidris alpina*), a small wader feeding on invertebrates. Poisoned birds contained c. 11 ppm of lead on average. Molluscs that they fed upon contained

c. 1 ppm of lead, suggesting that strong bioaccumulation had occurred. Smaller numbers of dead birds were found over the next two years. Estuaries are rich habitats for gulls, waders, and ducks, and this example illustrates the damage that can be caused to populations of birds by the careless disposal of chemical wastes.

ORGANOARSENIC COMPOUNDS

As with methylmercury compounds, methylarsenicals are both man-made and synthesizsed by microorganisms in the environment from inorganic forms. Methylarsenic compounds have been used as herbicides. One of these, dimethylarsenic acid (Agent Blue), was employed as a defoliant during the Vietnam War, an event that attracted adverse publicity in the media. Significant quantities of organoarsenic compounds have been found in marine organisms, and there have been concerns about the risks of consuming certain seafood, including canned tuna, on account of the content of organic arsenic.

Methylation of inorganic arsenic can lead to the formation of the highly toxic gas trimethylarsine, $As(CH_3)_3$ (see Crosby 1988). A classic example of this was the poisoning of humans by trimethylarsine produced by microbial action upon Paris Green, a form of inorganic arsenic and once a constituent of wallpaper.

SUMMARY

Organometallic compounds are formed when organic groups such as methyl or ethyl are bonded to metals or metalloids. The attachment of an organic group to metals or metalloids tends to modify their properties and environmental behavior; in particular, these organic groups tend to cause an increase in lipophilicity. Organometallic compounds or ions can readily cross membranous barriers such as the blood brain barrier, in contrast to the metal ions from which they are derived. This is why they can have toxic effects upon the brain.

Some organometallic ions found in the living environment (e.g., methylmercury and methylarsenic) are synthesized naturally by microorganisms as well as arising from chemical synthesis by man. Because of this, it can be difficult to determine the source of their residues in the environment.

Organomercury compounds were once widely used as seed dressings in Scandinavia, and were implicated in the secondary poisoning of predatory birds such as goshawks. Organotin compounds have caused marine pollution when used for descaling boats. Dog whelks and oysters have been affected by this practice. The disappearance of dog whelks from extensive stretches of the coast of southern England due to breeding failure illustrated an important point: that populations can decline, even to the point of localized extinction, due to a sublethal effect of a pollutant.

Organolead compounds used to be widely employed as an additive to petrol. This practice has gone into decline, because of the health risk to humans. There have been industrial incidents involving the poisoning of birds.

FURTHER READING

Alzieu, C. 2000. Impact of tributyl tin on marine invertebrates *Ecotoxicology* 9: 71–76.

Borg, K., Erne, K., Hanko, E., et al. 1970. Experimental secondary methyl mercury poisoning in the goshawk. *Environmental Pollution* 1: 91–104. An early publication giving evidence of bioaccumulation of methylmercury by a predatory bird.

Matthiessen, P., and Gibbs, P.E. 1998. Critical appraisal of the evidence for TBT-mediated endocrine disruption in molluscs. *Ecotoxicology and Environmental Safety* 30: 111–119. A concise review of the effects of TBT on molluscs.

Wolfe, M.F., Atkeson, T., Bowerman, W., et al. 2007. Wildlife indicators. In *Ecosystem response to mercury contamination: Indicators of change*, ed. R. Harris et al., 123–189. Boca Raton, FL: CRC Press, SETAC. A substantial review on the ecotoxicology of methylmercury.

Wolfe, M.F., Schwarzbach, S., and Sulaiman, R.A. 1998. The effects of mercury on wildlife: A comprehensive review. *Environmental Toxicology and Chemistry* 17: 146–160. A review on the effects of methylmercury on wildlife.

12 Pyrethroid and Neonicotinoid Insecticides

INTRODUCTION

These two groups of insecticides have a number of features in common, and there are advantages in describing them both in a single chapter. As has already been explained, both of them are derived from natural products that have insecticidal activity (Chapter 3). Pyrethroids are related to naturally occurring pyrethrins that are found in *Chrysanthemum* species. At one time pyrethrum, a preparation made from the flowering heads of these plants and containing natural pyrethrins, was used quite extensively as an insecticide. It came to be replaced by the analogous synthetic pyrethroid insecticides, which are more stable, both chemically and biochemically, than the natural product. Nicotine, another naturally occurring compound, was also once used as an insecticide. Commercially produced nicotine insecticide was extracted from tobacco plants (*Nicotiana tabacum*), which are rich in the chemical. Subsequently, the neonicotinoid insecticides were developed, which bear a structural resemblance to nicotine and express toxicity through the same mechanism. Like the synthetic pyrethroids, these are more effective insecticides than the natural products to which they are related.

Both groups of insecticides are neurotoxins and express their toxicity through the same mechanism as the related natural products. Thus, pyrethroids act upon sodium channels of the nerve membrane and thereby disturb the passage of impulses along nerves. Neonicotinoids act upon certain receptors for acetylcholine (nicotinic receptors) and thereby disturb the transmission of messages across cholinergic synapses. These mechanisms of toxicity have been described in Chapter 6.

Both groups of insecticides are less chemically stable and more biodegradable than the persistent organochlorine insecticides. Thus, they are seen as being more ecofriendly than DDT, dieldrin, and many other organochlorines. Pyrethroids came to replace these organochlorine insecticides for many uses. Neonicotinoids were developed later than pyrethroids, but they too came to be used for the control of pests that had formerly been controlled with organochlorines.

Finally, both groups have generally lower vertebrate toxicity than the earlier organophosphorous, organochlorine, and carbamate insecticides that were once widely used. They are seen as being less hazardous to use and more ecofriendly than many of their predecessors. However, the devil can be in the details, and problems have arisen with these compounds in the light of experience.

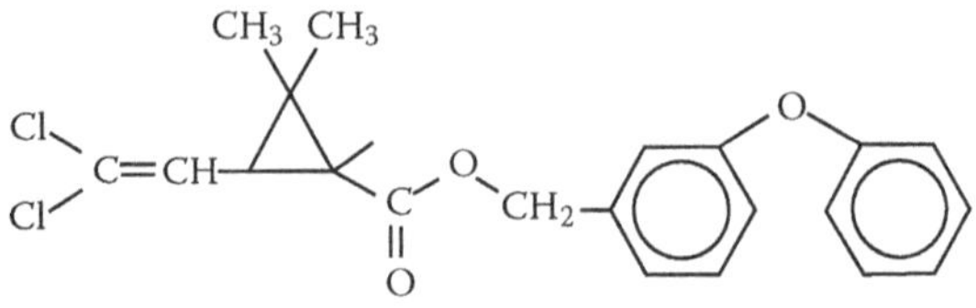

Permethrin

FIGURE 12.1 The structure of permethrin.

THE SYNTHETIC PYRETHROIDS

The synthetic pyrethroids came into wide use during the 1970s, often as replacements for older insecticides such as organochlorine, organophosphorous, and carbamate insecticides. Included among these were permethrin, cypermethrin, deltamethrin, and fenvalerate. In general, they are lipophilic solids of low water solubility. Chemically they are esters—consisting of an acidic component linked to a basic component. This is illustrated in Figure 12.1, which shows the structure of permethrin.

On the left-hand side of the formula is the acidic component, which is linked to the basic component on the right-hand side by an ester bond. The ester bond is C-O-C and can be broken by the attack of esterase enzymes, thus releasing the acidic and basic components as separate entities. This leads to a loss of toxicity. Permethrin can also be metabolized—and detoxified—by oxidative attack. Here, cytochrome P450-based oxidizing enzymes can introduce hydroxyl (OH) groups into the molecule.

The pyrethroids described here are not systemic. They have been widely employed to control phytophagous insects infesting crops. They have also been used for veterinary purposes, and for the control of mosquitoes and other vectors of disease. Some data on the toxicity of pyrethroids are given in Table 12.1.

There is evidence of marked selectivity here. Toxicity to the rat is not especially high—and borders on the boundary between high toxicity and medium toxicity given in the classification presented earlier (see Chapter 1). These indicate much lower toxicity than is shown by many organochlorine insecticides (OCs) or carbamate insecticides (note that higher lethal doses signify lower toxicities). By contrast, toxicities to birds are surprisingly low, with median lethal dose (LD_{50}) values running into the

TABLE 12.1
Toxicity of Some Pyrethroids

Compound	LD_{50} Rat mg/kg	LD_{50} Birds mg/kg	LC_{50} Fish at 96 h μg/L
Permethrin	500	13,000	0.6–314
Cypermethrin	250	10,000	0.4–2.8
Fenvalerate	451	4000	0.3–200

Source: Walker (2008).

thousands. From this point of view, the readily biodegradable pyrethroids appear to be relatively safe insecticides. However, this is not the case with fish or aquatic invertebrates. As can be seen, fish median lethal concentration (LC_{50}) values range from 0.3 to 314. LC_{50} values toward aquatic invertebrates range from 0.008 to 5. There are strict regulations in Western countries forbidding the release of pyrethroids into or near surface waters.

ENVIRONMENTAL FATE AND CONCERNS

Because of their lipophilicity and low water solubility, there might appear to be a high risk of biomagnification in food chains. However, pyrethroids are much more readily metabolized to water-soluble products (metabolites and conjugates) than are persistent pollutants such as dieldrin, p,p'-DDE, dioxins, polychlorinated biphenyls (PCBs), methylmercury, etc., and they do not tend to undergo biomagnification in food chains to the extent that these very persistent pollutants do.

As we have seen, fish tend to have low capacities for metabolizing organic pollutants in comparison with land animals and birds. Despite this, bioconcentration of pyrethroids from surface waters is low in fish and aquatic invertebrates in comparison with that found with very persistent pollutants. Bioconcentration factors for fish of 50 up to several thousand have been reported, whereas the values for p,p'-DDE and dieldrin nearly always run into thousands. We come back to the point that pyrethroids tend to be quite rapidly metabolized in comparison with these highly persistent compounds.

An important feature of pyrethroids is that they tend, like persistent organochlorine compounds, to be strongly adsorbed by colloidal particles (typically clay or organic matter) that are present in surface waters, sediments, and soils. This is because of their high lipophilicity and associated low water solubility. Adsorption implies strong binding to these particles and correspondingly low concentrations in water. Pyrethroids bound in this way can be quite persistent, but of only limited availability to aquatic organisms such as fish, invertebrates, and other denizens of the lower levels of aquatic food chains.

There is concern about potentially harmful effects of pyrethoids upon vulnerable aquatic organisms. There are strict regulations about the application of them in the vicinity of rivers, lakes, etc. In Western countries boundaries are usually set around water bodies within which pyrethroids must not be applied.

There has been controversy about the possible side effects of pyrethroids upon bees and other pollinators. Similar concerns have been expressed about certain neonicotinoids. This issue will be discussed for both types of insecticide in a later section.

THE DEVELOPMENT OF RESISTANCE TO PYRETHROIDS

With the worldwide use of pyrethroids over decades, examples of resistance in target pest species have come to light (see McCaffery 1998). As explained in Chapter 5, these have been of two types:

1. Resistance due to target insensitivity
2. Resistance due to enhanced detoxication

Mutant forms of a sodium channel or enhanced detoxication by cytochrome P450-based monoooxygenase have been found in resistant insects, and some resistant strains have possessed both of these resistance mechanisms.

A striking example of resistance to pyrethroids was found in the cotton fields of the southern United States during the latter part of the 20th century. Following many years of heavy use of pyrethroids, some strains of the tobacco bud worm, an important pest of cotton, were found to have developed such high levels of resistance that insecticidal control was becoming uneconomic. These levels were up to several hundred-fold in extreme cases. In these strains both of the above mechanisms were found to be operating. This episode well illustrates the dangers of overusing pesticides. Cotton, not being an edible crop, can be treated with relatively high levels of these insecticides. Edible crops are subject to strict limits in the levels of insecticides applied because of regulations on residue levels in food.

THE NEONICOTINOIDS

The natural product, nicotine, acts upon one type of acetylcholine receptors. That is why this type of receptor has come to be known as the nicotinic receptor for acetylcholine (see Chapter 6). The neonicotinoid insecticides have been modeled upon the structure of nicotine, and can interact with the nicotinic acetylcholine receptor. This process, which disrupts synaptic transmission, represents their mode of action. However, they differ in certain respects from nicotine itself. They are less water soluble and show greater toxicity to insects than to vertebrates. Nevertheless, they are water soluble enough to have systemic action and, in this respect, differ from the highly lipophilic pyrethroid insecticides described earlier.

Imidacloprid is an example of a neonicotinoid (Figure 12.2). Other examples include thiacloprid, clothianidin, acetamiprid, and thiamethoxam. All have moderate water solubility and have been successful in controlling sap-feeding insects such as aphids.

Neonicotinoids have quite variable toxicity to birds. Some examples are given in Table 12.2. Imidacloprid is highly toxic to grey partridge and house sparrow. Acetamiprid is highly toxic to the zebra finch. Questions have been raised about the risks of using seed dressed with imidacloprid and certain other neonicotinoids to birds (Mineau and Palmer 2013).

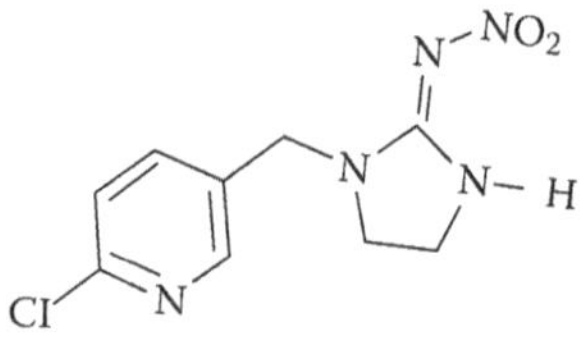

NO_2
N
N
N– H
Cl
N

FIGURE 12.2 The structure of imidacloprid.

TABLE 12.2
Acute Toxicities of Some Neonicotinoids to Birds

Neonicotinoid	Bird	Acute Oral LD_{50} mg/kg	Comment
Imidacloprid	Bobwhite quail	152	Effects seen from 50–100 mg/kg
Imidacloprid	Mallard	283	Severe effects from 25 mg/kg
Imidacloprid	Grey partridge	15	
Imidacloprid	House sparrow	41	
Acetamiprid	Bobwhite quail	180	
Acetamiprid	Mallard	98	Severe effects from 52 mg/kg
Acetamiprid	Zebra finch	5.7	Serious effects seen from 2.5–3.6 mg/kg

At the time of writing imidacloprid is one of the most widely used insecticides in the world. It is used as a seed dressing and a spray and has been successful in controlling pests of cereals, legumes, potatoes, and cotton.

DEVELOPMENT OF RESISTANCE TO NEONICOTINOIDS

It took some time for resistance to neonicotinoids to develop. However, there is now evidence of resistance to imidacloprid in peach-potato aphids, tobacco whitefly, damson-hop aphids, Colorado beetles, and other pest species (Jeschke and Nauen 2008). Often it has been associated with cross-resistance to other neonicotinoids. A serious example of resistance to neonicotinoids in a strain of the peach-potato aphid has been reported in southern France (Puinean et al. 2013). The mechanism of resistance was target insensitivity; this was due to the appearance of mutant forms of the nicotinic acetylcholine receptor with a low affinity for neonicotinoids. Another resistance mechanism operating in this highly resistant strain of the aphid was increased detoxication—due to enhanced cytochrome P450-based monooxygenase metabolism (see Chapter 5 for other examples).

The effects of neonicitinoids on bees will be discussed in a later section.

POTENTIATION OF THE TOXICITY OF PYRETHROIDS AND NEONICOTINOIDS

The questions of potentiation and synergism were discussed in Chapter 2 (see Box 2.1). Both pyrethroids and nicotinoids are readily metabolized (Walker 2009; Jeschke and Nauen 2008), and this, as we have seen, makes it unlikely that they will undergo the levels of biomagnification that have been found with p,p'-DDE, dieldrin, dioxins, or highly chlorinated PCBs. It does, however, raise the possibility that they may be subject to synergism in the presence of other chemicals that can inhibit enzymic detoxication.

Metabolism usually results in detoxication. The cytochrome P450-based monooxygenases have an important role in the metabolism of both groups of compounds.

A number of studies have shown that the toxicity of pyrethroids to bees can be synergized by inhibitors of these enzymes. These inhibitors are known to include piperonyl butoxide and a number of ergosterol biosynthesis inhibitor (EBI) fungicides (see Colin and Belzunces 1992; Pilling et al. 1995; Walker 2008). In these studies it was found that the presence of EBIs such as prochloraz could increase the toxicities of pyrethroids by 10- to 20-fold. Some incidents in France and Germany involving the poisoning of bees were attributed to the effects of tank mixes containing both pyrethroid insecticides and EBI fungicides. The EBIs were held to have had a synergistic effect upon the pyrethroids.

Synergism in honeybees has also been reported with neonicotinoids when combined with EBI fungicides (Iwasa et al. 2004; Schmuck et al. 2003). In one example, the toxicity of thiachloprid was reported to have increased by up to several hundredfold in the presence of an EBI (Iwasa et al. 2004). On present evidence imidacloprid does not appear to be subject to this type of synergism to any significant degree—probably because oxidative metabolism in this case can both increase and decrease toxicity. One metabolite is more insecticidal than the original insecticide (cf. the activation of organophosphorous insecticides (OPs) such as diazinon, malathion, and pirimiphos-methyl; Chapter 10).

EFFECTS OF PYRETHROIDS AND NEONICOTINOIDS ON HONEYBEES

There has long been concern about the lethal toxicity of insecticides to honeybees. The interest of beekeepers in this is understandable. However, there is a wider issue here as well. The effective pollination of some agricultural and horticultural crops depends upon the activity of honeybees and natural pollinators like bumblebees and hover flies. When highly toxic OPs such as triazophos were on the market, use of them was prohibited when certain crops such as oil seed rape were in flower—in order to protect foraging honeybees.

In recent times more ecofriendly insecticides such as pyrethroids and neonicotinoid insecticides have replaced earlier, more toxic, chemicals. For the most part, honeybees have not been exposed to lethally toxic levels of these newer insecticides in the field. That said, there have been incidents involving these compounds. Sometimes these have been due to misuse; e.g., doses used have been too high or inappropriate formulations of seed dressings have been used. Also, kills have been attributed to the mixing of pyrethroids with EBI fungicides in the spray tank.

There has been much concern about large-scale declines of honeybees over recent decades—declines that have not been convincingly explained. It seems very probable that there has been more than one cause. Insecticides have been suspected as being one of them. While there have been reports of substantial kills of bees associated with pesticide use, these do not explain the severe and widespread declines that have occurred.

A difficult question here is about sublethal effects of insecticides upon bees. With the systemic neonicotinoids, low levels of insecticide can be distributed throughout the plant and find their way into nectar and pollen—to be taken up by bees and other foraging insects. Further, both types of insecticide are neurotoxic, and it is well documented that they can have sublethal effects on bees, including effects upon behavior (Thompson 2003).

Regarding sublethal effects, it has been shown that a neonicotinoid can affect the wagtail dance of honeybees, which is used to transmit messages between individuals and indicate where nectar is to be found. This disturbance of transmission of information between individuals can disrupt foraging by the bees (see Schmuck 1999). Sublethal effects can also affect navigation when bees are attempting to return to the hive. This effect was recently reported in a study in France using the neonicotinoid thiamethoxam (Henry et al. 2012). Behavioral effects of a neonicotinoid have also been reported in bumblebees in the UK (Whitehorn et al. 2012). Also, when bumblebees were exposed to sublethal levels of imidacloprid, the colonies showed a reduced growth rate and an 85% reduction in the production of queens.

A number of studies relating sublethal doses of neonicotinoids known to occur in nectar or pollen under field conditions to sublethal effects upon honeybees have been inconclusive. These studies have looked at aspects of performance—either of individuals or of colonies. They have included aspects of growth, fecundity, longevity, or behavior. However, Cresswell (2011) subjected these results taken collectively to the statistical technique of meta-analysis and found that although there was no evidence of lethal effects at these levels of exposure, there was a reduction of between 6% and 20% in performance that was statistically significant.

This ongoing issue is of fundamental interest to ecotoxicologists. We are not dealing here with the readily apparent problem of acute lethal toxicity—as was found with major field incidents involving the poisoning of birds by organochlorine, organomercury, or organophosphorous pesticides in the second half of the twentieth century. We are now living in an era when—in developed countries at least—more ecofriendly pesticides are being used and there are questions about the possible ecological significance of sublethal effects that can be difficult to identify and quantify. Also, there is the possibility that interactive effects may occur between different pesticide residues in the field. There is a need for new approaches and techniques to address these questions in a scientific way. This issue will be returned to in Chapter 19.

SUMMARY

Pyrethroids and subsequently neonicotinoids came into prominence as insecticides during the latter part of the twentieth century. Both types of insecticide are neurotoxic. They are readily biodegradable, of limited water solubility, and have been regarded as more environmentally friendly than earlier organochlorine, organophosphorous, and carbamate insecticides, which they have come to replace. Unlike persistent organochlorine insecticides, they do not tend to be strongly biomagnified with movement along food chains. That said, some environmental problems with them have emerged.

Pyrethroids have low water solubility and tend to bind strongly to colloids such as clay minerals and soil organic polymers. This can limit bioavalability, but it can also lead to persistence in the environment. Pyrethroids are quite toxic to many aquatic organisms, and there are strict regulations in many countries to prevent the application of them in the immediate neighborhood of rivers, streams, and other surface waters. Unlike pyrethroids, neonicotinoids have moderate water solubility and show systemic properties.

There has been controversy about the possible effects of both types of insecticide upon bees and other pollinating insects. There is evidence that as neurotoxic compounds, they can have sublethal behavioral effects on bees that disturb communication between individuals. Such effects may affect the ability of individuals to locate nectar or find their way back to hives. This concern has been strengthened by evidence that their potency in the field may be increased by the synergistic effects of EBI fungicides that are also widely used as systemic agricultural pesticides.

Resistance has developed to both groups of insecticides in certain target species.

FURTHER READING

Cresswell, J.E. 2011. Meta analysis of experiments testing the effects of a neonicotinoid (imidacloprid) on honeybees. *Ecotoxicology* 20: 149–157. An analysis of published data on possible effects of a neonicotinoid insecticide upon bees.

Henry, M., Beguin, M., Requier, F., et al. 2012. A common pesticide decreases survival and foraging success in honey bees. *Science Express Report*, March 29, 2012, pp. 1–4. A field experiment giving evidence suggesting sublethal effects of a neonicotinoid on the behavior of bees.

Jeschke, P., and Nauen, R. 2008. Neonicotinoids from zero to hero. *Pest Management Science* 64: 1084–1098. A review of the neonicotinoids.

Leahey, J., ed. 1985. *The pyrethroid insecticides*. London: Taylor & Francis. A useful text on this group of insecticides.

Pilling, E.D., Bromley-Challenor, K.A.C., and Walker, C.H. 1995. Mechanism of synergism between the pyrethroid insecticide lambda-cyhalothrin and the imidazole fungicide prochloraz in the honeybee (*Apis mellifera* L.). *Pesticide Biochemistry and Physiology* 51: 1–11. Evidence for synergistic effects of an EBI on a pyrethroid in the honeybee.

13 PCBs and Dioxins

INTRODUCTION

During the 1960s, analysts who were detecting low levels of organochlorine insecticide residues in tissue samples from wild animals and birds by gas chromatography encountered a problem. Employing an electron capture detector, they often found a group of slow-running peaks that they were unable to identify. After a period of uncertainty, these peaks were found to be polychlorinated biphenyls (PCBs) (Jensen 1966). A later gas chromatogram obtained using the same detection system, but fitted with a high-resolution capillary column, is shown in Figure 13.1. It gives results obtained using extracts from biota of the Dutch Wadden Sea during the later 1980s (Boon et al. 1989).

The upper chromatogram represents a mussel, the lower one a seal. Each of the numbered peaks represents a single PCB congener. Following hard on the heels of the discovery of residues of the organochlorine insecticides (OCs) in the natural environment, this finding set alarm bells ringing. The PCB residues originated from a number of widely used industrial products employed as hydraulic fluids, coolants, insulation fluids in transformers, and plasticizers in paints. There was clear evidence for the biomagnification of some of these congeners in food chains. Were they having harmful effects on the natural environment?

After this discovery, another group of polychlorinated compounds were identified as pollutants. These were the dioxins. They were found to be present in biota at much lower concentrations than the PCBs, but certain of them were found to be highly toxic. The dioxins differ from the PCBs in that they are not synthesized for commercial purposes. They are unwanted by-products of the chemical industry. They are also by-products of paper bleaching and waste disposal.

Both PCBs and dioxins are stable fat-soluble compounds of low water solubility, and many of them tend to undergo biomagnification in food chains. As with the organochlorine insecticides, the persistent PCBs and dioxins have high levels of chlorination that make them resistant to enzymic metabolism.

Generally speaking, the most persistent congeners in the environment have the highest levels of chlorination. They have been included, together with certain organochlorine insecticides, among the "dirty dozen" pollutants termed persistent organic pollutants (POPs) by the United Nations Environment Program (UNEP 2011). Because of the similarities between PCBs and dioxins and the fact that they often coexist in the same ecosysem, they will be considered together in a single chapter.

THE POLYCHLORINATED BIPHENYLS (PCBS)

PCBs were first marketed in the 1930s. The commercial products composed of PCBs have been complex mixtures of different congeners (i.e., group members) with

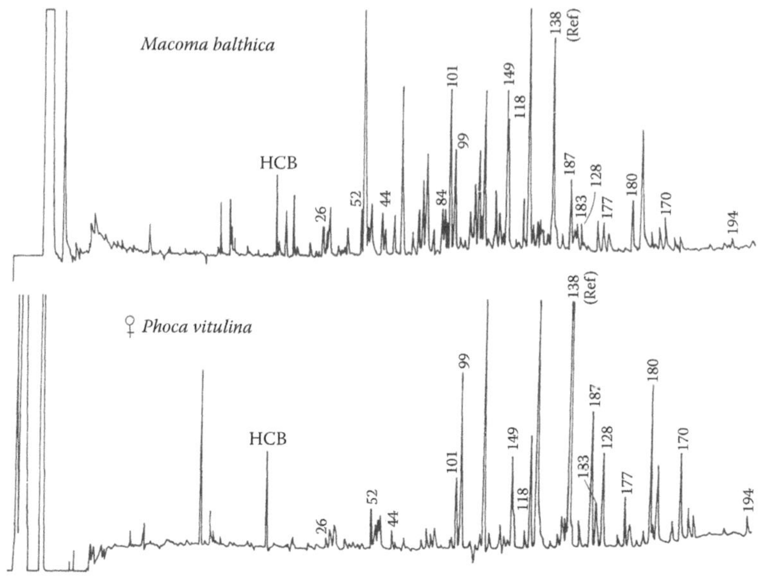

Gas chromatograms of extracts from *M. balthica* and *P. vitulina* (SE-54 capillary column). The PCB congeners are identified with their IUPAC numbers.

FIGURE 13.1 PCB congeners in tissues of marine organisms [mussels (*Macoma baltica*) and harbor seals (Phoca vitulina)] from the Dutch Wadden Sea. The compounds were separated, identified, and quantified by capillary gas chromatography. Each of the numbered peaks represents a PCB congener. HCB (hexachlorobenzene) served as an internal standard. (*Source:* Boon, J.P. et al. (1989). *Marine Environmental Research* 27, 159-176. With permission.)

names such as Aroclor 1242 or Aroclor 1260. The numbers given in these commercial names indicate levels of chlorination. Thus, Aroclor 1242 is 42% chlorinated, Aroclor 1260 is 60% chlorinated, and so forth. The structure of some PCB congeners is shown in Figure 13.2. In general, they are very chemically stable viscous liquids, of low electrical conductivity and low vapor pressure. These properties have made them suitable for a number of uses.

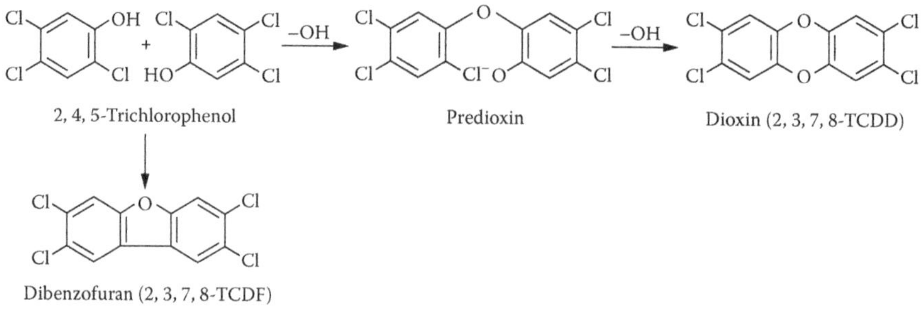

2, 4, 5-Trichlorophenol —OH→ Predioxin —OH→ Dioxin (2, 3, 7, 8-TCDD)

Dibenzofuran (2, 3, 7, 8-TCDF)

FIGURE 13.2 Some PCB congeners.

The major commercial applications of PCBs have been as follows:

1. As dielectrics in transformers and large capacitors
2. In heat transfer and hydraulic systems
3. In the formulation of lubricating and cutting oils
4. As plasticizers in paints
5. As ink solvents in carbonless copy paper

PCBs are synthesized by the action of chlorine upon biphenyl. In theory, there are no fewer than 219 possible congeners of PCB. However, only about 130 of these are likely to be found in commercial products. Looking at the basic structure, two rigid, flat benzene rings are joined together by a chemical bond. Chlorine atoms replace hydrogen atoms on the rings. The structures of three PCB congeners are shown in Figure 13.2. Working from left to right, 3,3′,4,4′-tetrachlorobiphenyl (3,3′,4,4′-TCB) has four substituted chlorine atoms, all in positions on the two benzene rings that are remote from the carbon-carbon linkage that binds them together. With no obstructing chlorines near to this linkage, the two benzene rings are naturally oriented in the same plane as one another in 3,3′4,4′-TCB. It is a flat coplanar PCB. The same description applies to the next structure, which is 3,3′,4,4′5,5′-hexachlorobiphenyl (3,3′,4,4′,5,5′-HCB). On the other hand, the third structure, that of 2,2′4,4′,6,6′-hexachlorobiphenyl (2,2′,4,4′6,6′-HCB), is not coplanar. There is steric hindrance between the bulky chlorines in the adjacent 4,4′ and 6,6′ positions, and the benzene rings are, so to speak, pushed out of their coplanar orientation. This steric hindrance between adjacent chlorines of different rings is not found in the first two structures, which are coplanar in structure. This distinction between coplanar and noncoplanar PCB congeners is important. Coplanar PCBs differ from other PCBs in their metabolism and in their toxicological properties.

Although PCBs, in general, tend to be chemically stable, many congeners are susceptible to attack by the cytochrome P450-based oxidative enzyme systems (see Box 2.1). This type of metabolism introduces hydroxyl groups (OH–) into the molecules, thus converting them into phenols. These chlorinated phenols are more water soluble than the original molecules. Chlorine substitution protects the molecule against this type of metabolic attack. Thus, lower chlorinated PCBs tend to be more readily metabolized than the more highly chlorinated ones. There is a difference here between coplanar PCBs and noncoplanar ones. Coplanar PCBs are metabolized by oxidative enzyme systems containing the cytochrome P450 form designated as cytochrome P450 1. This type of cytochrome P450 has a planar binding site to which planar liposoluble molecules such as polycyclic aromatic carcinogens (e.g., benzo(a)pyrene) and coplanar PCBs become strongly attached. By contrast, other PCBs that do not have this coplanar structure are metabolized by oxidative systems containing other forms of cytochrome P450. Cytochrome P450 2 is one of these other forms. It tends to metabolize substrates with a globular structure (see Lewis 1996; Walker 2009).

Returning to the question of biomagnification, the lower chlorinated PCBs present in the lower trophic levels of food chains are sometimes not found in the higher ones. This is illustrated in Figure 13.1, which contrasts the residues found in a mollusc belonging to a low level in the food chain with those found in a seal that occupies a higher position in the food chain. The PCB congeners with the higher levels of chlorination are to the right-hand side of the chromatogram and are assigned the highest numbers; those with the lowest levels of chlorination lie to the left-hand side and are assigned the lowest numbers. These numbers correspond to *relative retention times* on the analytical machine (the gas chromatograph). The relative retention times are related to the distances between the point of injection and the position of peaks on the gas chromatogram. In general, lower chlorinated PCBs are more volatile and move more quickly through the machine than do the higher chlorinated ones.

It is interesting to compare the ratios of the peak areas for higher chlorinated PCBs 170 and 180 to the peak areas of the lower chlorinated PCBs 52 and 101 in the two species. The peak areas provide a measure of the quantity of PCB being analyzed. The larger the area, the higher the quantity. With progression from the lower trophic level (the mollusc) to the higher trophic level (the seal), the peak areas for the higher chlorinated PCBs become larger in relation to those of the lower chlorinated PCBs. This happens because the higher chlorinated PCBs are being more strongly biomagnified than the lower chlorinated ones with movement along the food chain. A major factor determining this is evidently metabolism. The lower chlorinated PCBs are being more rapidly metabolized in the higher levels of the food chain than are the more highly chlorinated ones.

EARLY POLLUTION INCIDENTS INVOLVING PCBS

Box 13.1 describes a method that has been developed for estimating the combined toxicities of coplanar PCBs and dioxins. This technology was not available when pollution by these compounds was first discovered. Some early investigations of serious pollution incidents involving PCBs did not consider possible effects of dioxins. Often, data were available for dioxins because of the limitations of the analytical method used at that time. Even where there were some data for dioxins as well as PCBs, the approach described in this box was not followed. Some early incidents involving PCBs will now be described.

In the autumn of 1969 there was a large wreck of seabirds in the Irish Sea (Natural Environment Research Council 1971). A large number of dead birds, estimated to be over 17,000, was washed ashore on the coasts of Britain and Ireland. Most of these were guillemots, but there were also significant numbers of razorbills, gannets, and herring gulls (Walker and Livingstone 1992). All of these species feed upon fish or marine invertebrates.

At the time, the technique of analysis by gas chromatography was at an early stage of development. Separation of components of the extracted mixture was accomplished using ordinary packed columns that only gave limited separation of the different congeners. Different peaks overlapped one another, and it was not possible to make reliable estimates of the quantities of different congeners. There was not the clear separation of individual congeners that was possible on the more effective capillary columns that came into widespread use later on. Figure 13.1 illustrates the clean separation of different congeners that became possible using this improved technique.

BOX 13.1 A METHOD FOR ASSESSING THE COMBINED TOXICITY OF PCBS AND DIOXINS

The coplanar PCBs can express toxicity toward animals by a mechanism that is also operated by the dioxins. The mechanism is one that applies to certain *planar* molecules. In the cytosol of cells there is a receptor protein that binds aryl-hydrocarbons such as the carcinogen benzo(a)pyrene. This protein is termed the *aryl hydrocarbon receptor*, abbreviated as the Ah receptor. The binding of coplanar PCBs and dioxins to this receptor triggers a collection of toxic responses in the organism, a phenomenon that has been referred to by the rather cumbersome title *Ah receptor-mediated toxicity*.

The toxic mechanism is a complex one that is only partially understood, and it would not be appropriate to go into great detail here. In brief, coplanar PCBs and dioxins bind to the receptor protein to form a complex molecular structure. This complex then travels to the nucleus of the cell where it interacts with DNA. These events lead to the appearance of a number of toxic effects, including dermal toxicity, immunotoxicity, and endocrine disturbances. This interaction with the Ah receptor by coplanar PCBs and dioxins also leads to the induction of cytochrome P450 1—a signal that indicates when toxic effects are taking place.

The important issue here is that a considerable number of dioxins and flat PCBs can interact with the same receptor, and when they do, there are resultant toxic effects. Moreover, these effects appear to be additive. Thus, there is a need to estimate the additive effect when a number of pollutants interact with the Ah receptor at the same time. This is done by estimating *dioxin equivalents* for each of the pollutants, then adding these together to obtain an estimation of overall *dioxin equivalence*. Overall dioxin equivalence is an estimate of the concentration of dioxin that would express the same toxicity as the entire mixture of PCBs and dioxins found in an environmental sample, e.g., animal tissue or bird's egg.

In early investigations of PCB pollution, the technology was not sufficiently advanced to adopt the above approach. Only limited analytical data were available. In the present text early investigations of (1) PCB pollution and (2) dioxin pollution will be dealt with separately. Later in the text, a further section will be devoted to more recent investigations that have utilized dioxin equivalents to enable an estimation of the overall toxic effect of both types of pollutant taken together.

***Source:* Walker (2009). *Organic Pollutants: an Ecotoxicological Perspective*, 2nd Ed.**

In the event, a crude estimate of total PCBs was made, using a PCB mixture as a standard. A basic limitation here is that some congeners are more responsive to detection than others—and, for reasons already given, the composition of the mixture found in biota at the top of the marine food chain tends to be significantly different from the composition of a commercial PCB mixture. Lower chlorinated PCBs are much less well represented than highly chlorinated ones in top levels of the food chain. Bearing this limitation in mind, estimated total PCB concentration in the livers of birds found dead ranged from 1 to 880 ppm.

Somewhat later, in 1970 and 1971, cormorants were found dead in incidents that were investigated in the Netherlands (Koeman and Pennings 1970; Koeman et al. 1973). These birds were estimated to contain high levels of PCB in liver—comparable to the levels in cormorants that had been experimentally poisoned (210–285 ppm). It was concluded that all or most of the birds had been poisoned by PCBs. Returning to the Irish Sea bird incident, many of the guillemots contained comparable PCB levels (i.e., 210 ppm or more) in the liver, and it seems probable that they died of PCB poisoning, despite the cautious wording of the report by the Natural Environment Research Council.

THE DIOXINS

The term *dioxin* is commonly applied to members of a group of highly chlorinated planar compounds known to chemists as polychlorinated dibenzodioxins (PCDDs). The structure of the most important of them, from a toxicological point of view, often referred to simply as dioxin, is shown in Figure 13.3.

The figure shows that it is formed by the interaction between two molecules of a trichlorophenol, which leads to the creation of a planar molecule consisting of two benzene rings linked together by two oxygen bridges. Referring to Figure 13.3, it can be seen that dioxin bears some resemblance to coplanar PCBs—except that it is more widely spaced because of the two oxygen bridges that create what is termed a dioxin ring.

Dioxins have not been deliberately synthesized by the chemical industry. They are unintended by-products. They are also formed by the combustion of wastes containing PCBs, and they are products of the paper industry, when wood pulp is treated with chlorine. In all cases, it comes back to the type of reaction shown in Figure 13.3 where chlorinated phenols interact with one another to form dioxins.

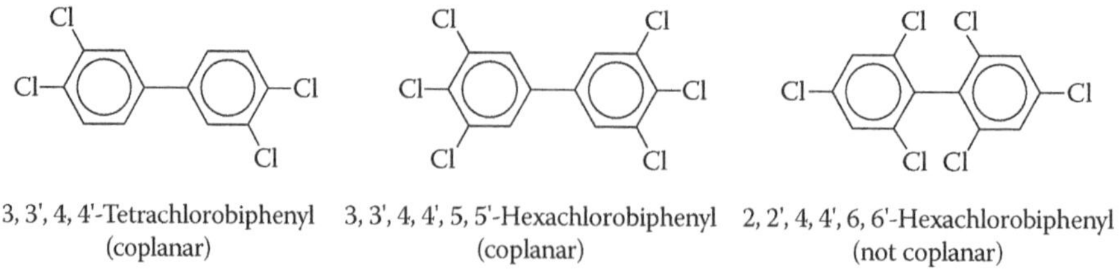

3, 3', 4, 4'-Tetrachlorobiphenyl (coplanar)

3, 3', 4, 4', 5, 5'-Hexachlorobiphenyl (coplanar)

2, 2', 4, 4', 6, 6'-Hexachlorobiphenyl (not coplanar)

FIGURE 13.3 The formation of dioxin.

One notorious case of dioxin pollution, which affected human beings as well as the natural environment, occurred during the Vietnam War. The defoliant Agent Orange was applied aerially to areas of jungle. The main ingredient of this preparation was 2,4,5-T, a plant growth regulator herbicide (see Chapter 14), which contained appreciable quantities of dioxin as a by-product. (Trichlorophenols are ingredients used in the synthesis of this type of herbicide.) One effect upon exposed humans was the development of skin acne. Following this incident, the synthesis of 2,4,5-T and related herbicides was more carefully regulated to ensure that levels of dioxin in the commercial product were kept very low.

Another case of dioxin pollution that was widely publicized occurred at a chemical factory at Seveso, Italy, where trichlorophenol antiseptic was being manufactured. Following an explosion, a cloud containing chlorinated phenols and dioxins was released that caused severe pollution of neighboring areas. People who were seriously exposed developed chloracne.

Dioxins have also been detected in the effluent of paper mills in North America, Scandinavia, and Russia. This is apparently due to the formation of chlorinated phenols when chlorine interacts with the lignin of wood pulp. Once again, chlorinated phenols interact to form dioxins.

The remainder of this section will be devoted to a description of the properties of 2,3,7,8-tetrachlorodibenzodioxin (see Figure 13.3), the most toxic of these compounds, which will be referred to simply as dioxin and taken as a representative of the group.

Dioxin is a stable solid of high lipophilicity and very low water solubility (0.01–0.2 μg/L). It is persistent in vertebrates showing biological half-lives of ten–ninety-four days in rodents.

Like higher chlorinated PCBs, it tends to undergo biomagnification in food chains, reaching relatively high concentrations in predators of the highest trophic levels. However, the levels found in biota are low in comparison to those of persistent PCBs. That said, these low levels are still of concern to environmentalists because dioxin is highly toxic to certain mammals. This is believed to be due, largely or entirely, to Ah receptor-mediated toxicity (Box 13.1).

THE COMBINED TOXICITY OF PCBS AND DIOXINS

The coplanar PCBs can express toxicity toward animals by Ah receptor-mediated toxicity, a mechanism that is also operated by the dioxins (Box 13.1). Molecules that can act in this way are planar in structure. Nonplanar (e.g., globular) molecules do not interact effectively with the site of action that is termed the Ah receptor. The binding of coplanar PCBs and dioxins to this receptor triggers toxic responses in living organisms. These two classes of pollutant are examples of polyhalogenated aromatic hydrocarbons (PHAHs).

The presence of dioxins in biota was not confirmed until the early 1980s because analytical techniques were not powerful enough to determine small quantities of these compounds in the presence of residues of PCBs and other organochlorine compounds. When technology improved, it became possible not only to determine concentrations of the dioxins (and certain related compounds termed PCDFs), but also to estimate dioxin equivalents for compounds known to express Ah receptor-mediated

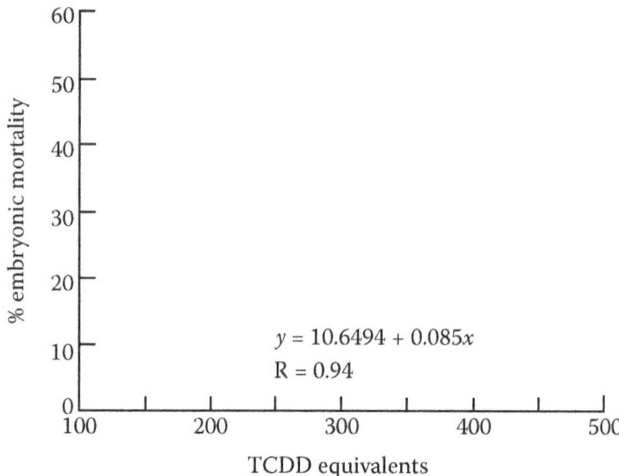

FIGURE 13.4 Relationship between dioxin equivalents and reproductive success in Caspian terns on the North American Great Lakes. (*Source:* Peakall, D. B. (1992). *Animal Biomarkers as Pollution Indicators*. Chapman & Hall. With permission.)

toxicity. The combined value of dioxin equivalents gave an estimate of the total Ah receptor-mediated toxicity expressed by a mixture of PCBs and dioxins. (The PCDFs are not discussed further here because they only made a minor contribution to the estimated dioxin equivalents in most ecotoxicological studies, and inclusion of them would make the present account unnecessarily complicated.)

An example of the application of this approach comes from a wide-ranging study of the complex pollution of the Great Lakes (Peakall 1992; Gilbertson et al. 1998). Reproductive failure was found in both double-crested cormorants and Caspian terns on Lakes Ontario and Michigan, and in both cases this correlation was reported between reproductive failure and the dioxin equivalents in the eggs of the birds determined by analysis. An example from this study is given in Figure 13.4. It can be seen that there is a good correlation between the dioxin equivalents and the embryo mortality in the eggs of these birds. In this example, over 90% of the dioxin equivalents were accounted for by two coplanar PCBs. A similar observation was made with the double-crested cormorant in the same Great Lakes study.

In later studies, a bioassay termed the Calux system was used to determine dioxin equivalents (see Chapter 2). This is a cellular preparation that includes the Ah receptor. When coplanar PCBs or dioxins combine with this receptor, light is emitted; the quantity of emitted light is measured, and this provides an estimate of the number of dioxin equivalents in an environmental sample. For further explanation of this approach, see Walker (2009).

In some studies of pollution by PCBs and dioxins, estimation of dioxin equivalents in environmental samples has been carried out by two methods: chemical analysis and bioassay. Correlations have then been sought between the estimated dioxin equivalents and biological effects, e.g., on reproductive parameters such as embryonic mortality. Sometimes the two approaches have given similar results, and sometimes there have been anomalies. For example, in some studies bioassays have suggested the presence of higher levels of PHAHs than have been found by chemical analysis (Walker 2009). This might suggest that there are compounds in certain samples that interact with the Ah receptor that are not detected by the method of chemical analysis that is employed. Readers are referred to recent literature for the latest evidence on this question.

SUMMARY

Following the discovery of residues of persistent organochlorine insecticides in environmental samples, other types of polyhalogenated pollutants with long biological half-lives have come to light. Prominent among these have been the PCBs and the PCDDs. The latter group are sometimes referred to simply as dioxins. Certain PCBs and dioxins undergo marked biomagnification with movement along food chains.

PCBs, an abbreviation for polychlorinated biphenyls, are industrial chemicals. First marketed in the 1930s, they have had a number of commercial applications, including as dielectrics, as components of commercial oils, and as plasticizers in paints. By contrast, dioxins are not commercial products but unintended by-products of manufacturing processes, papermaking, and waste disposal. Certain PCBs and dioxins have appreciable toxicity to vertebrate animals. Both coplanar PCBs and dioxins can express what has been termed Ah receptor-mediated toxicity. Methods have been developed to estimate the toxicity of mixtures of PCBs and dioxins found in environmental samples, and these have included bioassay systems such as Calux.

PCBs and dioxins have been implicated in investigations of environmental pollution. These have included kills of seabirds in Great Britain and the Netherlands and population declines of piscivorous birds in the United States and Canada.

FURTHER READING

Gilbertson, M., Fox, G.A., and Bowerman, W.W., eds. 1998. *Trends in levels and effects of persistent toxic substances in the Great Lakes*. Dordrecht: Kluwer Academic Publishers. Contains a number of reviews reporting results from a long-term study of persistent pollutants in the Great Lakes of North America. Includes much data on PCBs and dioxins.

Natural Environment Research Council. 1971. *The sea bird wreck in the Irish Sea*. Autumn 1969 NERC Publications Series C, No. 4. An official report of a large-scale seabird wreck that involved PCBs.

Walker, C.H., and Livingstone, D.R., eds. 1992. *Persistent pollutants in the marine environment*. Special Publication of SETAC. Oxford: Pergamon Press. Contains a number of reviews describing PCB and dioxin pollution of the marine environment.

14 Herbicides

INTRODUCTION

Herbicides are among the most extensively used of pesticides. They are applied to large areas of arable land—and also to parks and gardens in urban areas. Most have very low toxicity toward animals, and a fair number of them are readily available to the general public for weed control on lawns, vegetable plots, paths, and drives. This contrasts with the situation regarding insecticides, many of which have appreciable mammalian toxicity and are considered to be unsafe for general use. Only a very limited range of insecticides are generally available for use in gardens, etc.

Some important types of herbicide are identified in Table 14.1, and the structures of some of them in Figure 14.1. As can be seen, there are considerable differences in structure between the different types of herbicides illustrated here.

Often, any one type of herbicide will give effective control of only a limited range of weeds. Because of this, commercial formulations often contain two or more different herbicides to widen the spectrum of weeds that are controlled.

Why are many herbicides highly selective between plants and animals? A complete answer to this cannot be given, although it can be said that biochemical and physiological differences between animals and plants can provide the basis for selectivity. One important reason is that many herbicides act upon sites that are found in plants but not in animals. For example, the urea, triazine, uracil, and pyridazinone herbicides all act upon the photosynthetic system that is found in green plants but not in animals (Table 14.1). In general, their toxicity to animals is low.

Widely used herbicides such as methoxychlorophenoxyacetic acid (MCPA), methoxychlorophenoxybutyric acid (MCPB), 2,4-D, mecoprop, and dichlorprop all act in a similar way to naturally occurring auxins. They have been termed plant growth regulator herbicides and cause disordered growth and finally death when applied to plants (Hassall 1990). People who treat lawns with herbicidal mixtures containing these compounds will be familiar with the contorted growth of broad-leaved weeds that they cause.

Plants, unlike most insects and other animal pests, are immobile and consequently make "sitting targets" for the pesticides used against them. Unsurprisingly, herbicides have been successfully employed for controlling weed species from large areas of agricultural land. That said, plants, like insects, can develop resistance when the pesticides used to control them are overused.

Speaking generally, herbicides have caused less concern about their environmental side effects than have insecticides or rodenticides. Much attention has been given to the unintended effects of insecticides and rodenticides upon wild animals and birds, or to bees and other beneficial organisms. By contrast, there has been less concern about the unintended effects of herbicides on wild plants. This seems a little unbalanced. After all, the continuation of life itself depends upon the

TABLE 14.1
Some Important Herbicides

Type	Examples	Mode of Action	Uses	Comment
Plant growth regulator	MCPA, MCPB, 2,4-D, 2,4,5-T, mecoprop, dichlorprop	Auxin-like; disturb growth regulation	Control of dicots in monocot crops	Widely used to control weeds in cereal crops and on lawns
Ureas	Diuron, linuron, fenuron	Inhibit photosynthesis	Soil-acting herbicides	Widely used to control weeds in crops
Triazines	Atrazine, simazine	Inhibit photosynthesis	Soil-acting herbicides	Rather persistent compounds
Bipyridyl	Paraquat, diquat	Interact with photosynthetic system and generate destructive oxyradicals	Total weed killers; used for many purposes	Inactivated in soil because of binding to clay minerals; toxicity to mammals, including humans
Benzoic acids	Dicamba, TBA	Act as plant growth regulators	Control of dicots in monocot crops and lawns	Often used in mixtures
Only one example	Glyphosate	Disturbs amino acid metabolism	Total weed control; widely used	Root acting; inactivated in soil by binding to clays
Carbamates	Chlorpropham, barban	Chlorpropham disturbs cell division	Chlorpropham is soil acting; barban acts upon foliage	

Source: Hassall, K.A., *The Biochemistry and Uses of Pesticides*, Basingstoke: Hants Macmillan, 1990; Ashton, F.M., and Crafts, A.S., *Mode of Action of Herbicides*, New York: John Wiley, 1973.

photosynthesizing plants that occupy lowly positions in food chains. In ecotoxicology the primary concern is about effects upon populations of living organisms—and it follows that severe damage to plant populations is liable to have knock-on effects upon animal populations higher in the food chain.

In the following account this issue will be addressed before considering the direct effects of a few herbicides upon animals, the problem of herbicide misuse, the pollution of surface waters, and the conservation of rare plants on agricultural land.

THE IMPACT OF HERBICIDES ON AGRICULTURAL ECOSYSTEMS

Substantial areas of the globe are devoted to the growing of crops. Since the end of World War II herbicides have come to be very widely used to control weeds growing

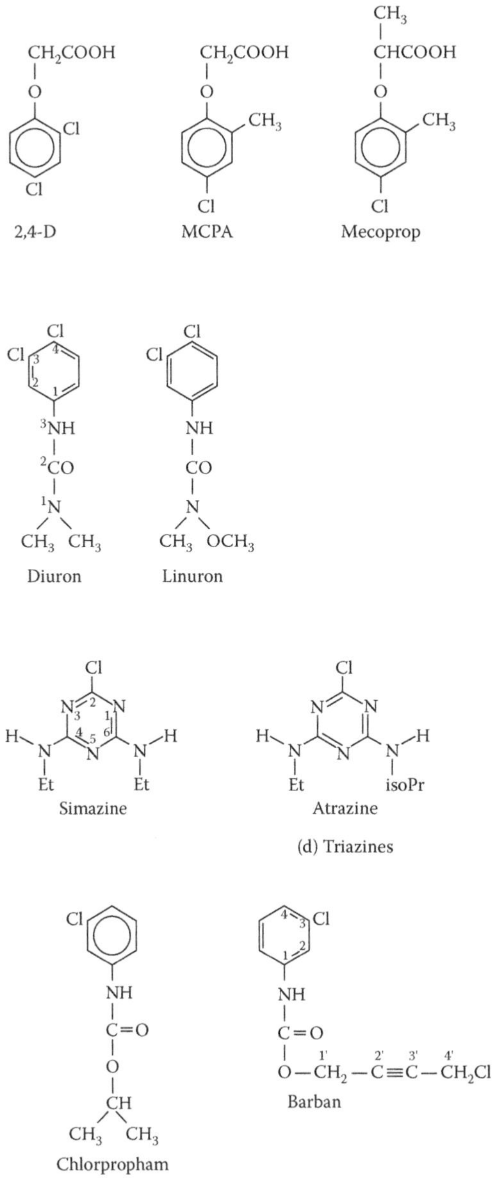

2,4-D MCPA Mecoprop

Diuron Linuron

Simazine Atrazine

(d) Triazines

Chlorpropham Barban

FIGURE 14.1 The chemical structures of some herbicides.

in such areas, especially in the developed world. The effectiveness of this practice, together with intensive cultivation of the land, has become apparent in agricultural areas of North America and Western Europe, where few weeds are seen. Weeds have been effectively controlled over substantial areas.

There has been growing concern about the effect of this drastic reduction of weed species upon agricultural ecosystems. In the UK ornithologists have noticed the decline of avian species on agricultural land in recent decades. Evidence for such declines has come from surveys conducted by the British Trust for Ornithology (BTO) and the Royal Society for the Protection of Birds (RSPB) (Campbell et al. 1997). One example has been the decline of the grey partridge. A study of the decline of this species in the UK was conducted by the Game Conservancy, commencing in the late 1960s (Potts 1986, 2000; Walker et al. 2012). In a study area in southern England, it was found that the decline of the grey partridge on agricultural land was related to a very high level of chick mortality. The high chick mortality was largely explained in this case by a shortage of their insect food, especially sawflies. The shortage of insects was related to the absence of the weeds upon which these insects feed. The use of insecticides appeared to make only a minor contribution to this shortage of insect food. Subsequent studies in southern England showed that grey partridge populations continued to survive if unsprayed headlands were established on the farmland habitat. Overall, the key factor in the decline of this species in this area was identified as a shortage of insect food for chicks—and the main reason for this shortage was the absence of weed species able to support a sufficient supply of insect food. This appears to be a prime example of an indirect effect of pesticides on an avian population. The herbicides in question only have very low avian toxicity and would have been regarded as safe to birds following statutory environmental risk assessment. For further discussion of this issue see Chapter 18.

During the latter part of the twentieth century other species of farmland birds also declined in the UK in areas where there was intensive arable farming (Campbell et al. 1997; Crick et al. 1998; Van Emden and Rothschild 2004, chap. 3; Walker et al. 2012, chap. 12). Species showing marked decline included tree sparrow, turtle dove, spotted flycatcher, and skylark. Inevitably, the question was raised about the possible role of pesticides in these declines. Not an easy question to answer since other factors, such as agricultural practice, habitat destruction, and climate change, might have been responsible. With migrant species such as the turtle dove and spotted flycatcher, declines might have been caused by events in their wintering areas of Southern Europe or Africa rather than in the UK during the breeding season.

Of the species mentioned above, the decline of the spotted flycatcher began as early as 1969, but the other three began to decline during the period 1978–1981. Campbell et al. (1997) showed that the initiation of these declines coincided with the increased use of pesticides on arable land. Some strong associations were found between the initiation of the declines and the extent to which cereal crops were treated with herbicides. The use of herbicides in cereals steadily increased over the period 1975–1998. In the case of the tree sparrow, turtle dove, and skylark the declines started when herbicides came to be applied to more than 50% of the cereal crop. A similar trend for these species was also found for application of fungicides to cereals, but not for insecticides or molluscicides in the same crop.

With the recognition of a serious decline in a number of species of farmland birds across Western Europe, measures have been implemented to try to reverse this trend. These have included not only limitations of pesticide use, but also changes in farm management intended to create more favorable habitats for birds and other wildlife. Included here have been the practices of "set aside"—leaving certain fields as fallow over a period of time, leaving stubbles over the winter when growing cereal crops, and leaving unsprayed headlands on cultivated fields. Within the European Union funds have been made available to farmers to support environmental stewardship schemes intended to promote more ecofriendly farming practices as part of the Common Agricultural Policy (CAP). With recent economic problems, less money has been made available for such purposes. The practice of set aside was discontinued in 2008, and the whole issue of environmental stewardship will be looked at in a review of CAP in 2013.

What has been the outcome of these ecofriendly practices in the UK? In the case of farmland birds the overall effect seems to have been rather disappointing. The best that can be said at the time of writing is that the rate of decline of certain species of farmland bird has slowed down, but there is no evidence of dramatic recovery (Annual Review of the British Trust for Ornithology 2011). That said, a clearer picture should emerge with the publication of surveys for the period in question.

Up to this point the focus of the discussion has been on the consequences of the control of weeds on agricultural land by herbicides. Weeds are usually regarded as being wild plants growing in the wrong place. From an agricultural point of view, they are wild plants that can effectively compete with crops, and so reduce yield. However, there are some plants once common on agricultural land that do not appear to have been weeds in this sense. Botanists have noticed the disappearance of interesting and attractive wild flowers from farmland with the widespread use of herbicides. In the UK, these have included the corn cockle (*Agrostemma githago*) and pheasant's eye (*Adonis annua*). It may be hoped that more ecofriendly practices will allow species like these to survive on farmland.

THE AERIAL MOVEMENT OF HERBICIDES

When herbicides are applied in the field, the intention is that they should remain within a particular area, and not contaminate neighboring areas. Fulfilling this objective is not always as simple as it may seem. Drifting herbicide has sometimes caused serious damage to crops grown in fields adjacent to the target area. When application is of sprays or dusts, there is the question of drift. Spray droplets or dusts can be carried over considerable distances if there is a strong wind. Thus, application should only be carried out when there is little or no wind. Even under calm atmospheric conditions there can be difficulties. With aerial spraying, for example, it can be very difficult to deposit all of the herbicide within the target area—even when there is limited air movement.

Drift is particularly a problem when an herbicide is volatile. Here, especially under warm conditions, significant quantities may vaporize when applied in the field. In the vapor form, an herbicide can diffuse into neighboring fields causing crop damage. This problem was encountered when plant growth regulator herbicides such as MCPA were formulated as esters. Early ester formulations had significant vapor pressures, and there was a problem of vapor drift when using them. This has not been a characteristic of other formulations of MCPA, etc., e.g., as sodium or potassium salts that are not particularly volatile and continue to be used.

Drift of herbicide is apparent when damage is caused to plants outside of the treated area. Usually the herbicide responsible can be identified by chemical analysis of the affected vegetation or surrounding soil. However, this was not always the case when the highly effective sulfonyl urea herbicides were first introduced. The herbicides were so effective that their residues were not detected by analytical methods available at the time.

The problem of drift applies not just to herbicides, but to pesticides more generally, and is important in the protection of natural habitats. As discussed in Chapter 12, pyrethroid insecticides are highly toxic to aquatic invertebrates and fish, and in some countries their use is strictly controlled. If applied near to surface waters, e.g., rivers and streams, exclusion zones may be defined within which application of pyrethroids is forbidden to protect neighboring freshwater ecosystems.

CONTAMINATION OF SURFACE WATERS WITH HERBICIDES

Herbicides have been extensively used over large areas of arable land in Western countries. On the face of things, there appears to be a serious risk of pollution of neighboring surface waters. After all, drainage water from agricultural land finds its way into water courses that are located around the fields, and these drain into rivers, lakes, and eventually the sea. However, this is not such a serious problem as it may seem. Many herbicides are quite persistent in soil, and tend to bind strongly to colloidal clay minerals and organic matter (humus). In the extreme case, some herbicides such as diquat, paraquat, and glyphosate are bound so strongly that they are effectively immobilized and cannot express herbicidal activity once they are incorporated in soil. Others, e.g., the urea and triazine herbicides, have low water solubility, are adsorbed strongly by soil colloids, and usually percolate only very slowly through the soil profile. More water-soluble herbicides such as MCPA, 2,4-D, and mecoprop show less tendency to bind to soil colloids, but are usually readily metabolized by soil microorganisms. So, under normal conditions, herbicides only find their way into drainage waters to a limited extent.

However, there are complications. One of these is that with heavy rainfall and consequent flooding, soil minerals can be washed from the surface into water courses, carrying attached herbicide with them. Some free herbicide may experience the same fate. Another problem is more localized. Where soils are high in clay minerals, they can develop cracks and fissures during dry weather due to shrinkage. If heavy rain follows, free herbicides can be washed down through the soil and into drainage water, without having to pass down through the soil profile where they would normally be bound to colloids or metabolized. Under these conditions, herbicides—and other pesticides—can appear in significant concentrations in drainage waters. The operation of this process has been identified in an agricultural area with clay soils subject to heavy bursts of rainfall (Williams et al. 1996). Herbicides reaching significant levels in drainage water in this study included atrazine, simazine, and MCPA.

A study of herbicide levels in some British rivers by House et al. (1997) showed that there were sporadic high levels of diuron and mecoprop, presumably arising as a result of periods of heavy rainfall. Sometimes these were high enough to be toxic to aquatic plants, but they were not sustained.

The levels of the triazine herbicide atrazine in drinking water have sometimes been high enough to cause concern about public health hazards. This has been reported in both North America and Western Europe.

HERBICIDES THAT HAVE SIGNIFICANT TOXICITY TOWARD ANIMALS

Most herbicides show only low toxicity to animals. In the main, they act upon sites within plants that do not exist in animals. For example, sites found within chloroplasts that are involved in the mediation of photosynthesis (e.g., ureas and triazines) or of hormonal regulation (e.g., MCPA, 2,4-D, mecoprop). However, there are a few exceptions to this generalization that deserve mention.

One is the herbicide 2,4,5-T. This is a plant growth regulator herbicide that was once used in the defoliant Agent Orange, and has already been described in Chapter 13. The existence of a plant growth regulator herbicide with significant mammalian toxicity may seem to run contrary to the main thrust of the previous paragraph. However, the critical point is that the mammalian toxicity is not due to 2,4,5-T but to the contaminant dioxin. The toxicity of dioxin is described in the previous chapter and will not be discussed further here. Suffice it to say that environmental toxicity is sometimes the result of the existence of impurities in industrial products—not the industrial products themselves.

An early herbicide with high mammalian toxicity is dinoseb. This is structurally related to dinitrophenol, a chemical long known to be highly poisonous to man. Dinoseb was once used to control weeds, but was banned in Western countries during the latter part of the last century because of its toxicity to humans. It is worth mentioning this because both dinoseb and the related dinitrophenol have an interesting mode of toxic action. They can act upon the mitochondria of both animals and plants.

Mitochondria have a vital role in energy conversion. Energy released during the breakdown of carbohydrates and fats can be stored in the form of an ionic gradient (a gradient of protons) across the inner membrane of the mitochondrion. Normally, this energy can be used to synthesize the vital molecule adenosine triphosphate (ATP), which is utilized to drive biochemical processes. These poisons are able to run down this gradient with consequent loss of energy as heat. This destructive process is described by biochemists as uncoupling of oxidative phosphorylation. It is a process common to both higher plants and animals—and this is why dinoseb is very toxic to both weeds and animals. More can be read about the uncoupling of oxidative phosphorylation in Harrison and Lunt (1980).

The bipyridyl herbicides paraquat and diquat are further examples of herbicides of significant mammalian toxicity. They have been implicated in poisoning incidents of hares and other lagomorphs on agricultural land (Sheffield et al. 2001). The hazards associated with paraquat were widely publicized following lethal poisoning incidents involving children. These resulted from the accidental consumption of the herbicide after it was stored in lemonade bottles. In plants paraquat and diquat interact with the photosynthetic system. This interaction leads to the generation of highly reactive oxyradicals, which can cause cellular damage. When animals are poisoned by these compounds, oxyradicals are also generated, although not, of course, following interaction with the photosynthetic system, which they do not possess. The underlying problem is that the bipyridyl compounds have an appetite for electrons, which they can obtain from electron transport systems of both plants and animals. In due course these electrons are passed on to molecular oxygen to form oxyradicals.

In animals, electrons come from sources other than the photosynthetic system. In the end, cellular damage appears to be caused not by the original herbicides, but by the active oxyradicals that they form following their acquisition of electrons inside either plants or animals (see Walker 2009).

Like glyphosate, paraquat and diquat are immobilized in soil and consequently lose their activity. They have been used as chemical plows, causing total weed control through action on foliage—but then expressing no activity once taken up by soil. They are positively charged molecules and, like the ammonium ion (NH_4^+), bind to negatively charged sites within the lattices of clay minerals.

SUMMARY

Most herbicides are not very toxic to animals. One reason for this is that their sites of action in plants (for example, in the photosynthetic system) do not exist in animals. There are a few exceptions to this generalization, notably paraquat and related herbicides and dinoseb. Agent Orange, a defoliant applied to the jungle during the war in Vietnam, was an example of an herbicide of significant mammalian toxicity, but this was due to the presence of dioxin as an impurity.

Herbicides have been very widely used to control agricultural weeds—with considerable success in many developed countries. This trend has contributed to a reduction in biological diversity on intensively farmed land. There is evidence of the decline of certain farmland birds that has been coincident with increase in the use of herbicides. In one study in the UK, the decline of the grey partridge was linked to the loss of weed species from agricultural land. The loss of weeds led to the reduction of insects feeding upon them—and consequently insufficient insect food for partridge chicks. This trend was shown to be reversed if field headlands were left unsprayed to retain some weeds—and therefore, some insect food for the chicks.

When herbicides are widely used, residues of them are found in surface waters. This tendency is marked in areas of the UK with heavy clay soils—due to the development of fissures through which herbicides can be eluted when there is heavy rainfall.

There have been problems with spray drift of herbicides that has led to damage to crops on neighboring land.

FURTHER READING

Campbell, L.H., Avery, M.I., Donald, P., et al. 1997. *A review of the indirect effects of pesticides on birds*. JNCC Report 227. Peterborough, UK: Joint Nature Conservation Committee. A report on a survey of farmland birds linking certain declines to the usage of pesticides.

Hassall, K.A. 1990. *The biochemistry and uses of pesticides*. Basingstoke: Hants Macmillan. A textbook that describes major classes of herbicides and their properties.

Potts, G.R. 1986. *The partridge*. London: Harper Collins. Describes the decline of the grey partridge and the role of herbicides in this decline.

Williams, R.J., Brooke, D.N., Clare, R.W., et al. 1996. *Rosemount Pesticide Transport Studies 1987–1993*. Report 129. Wallingford, UK: Institute of Hydrology (NERC). Explains the problem of herbicides contaminating surface waters in areas with heavy soils.

15 Endocrine Disruptors

INTRODUCTION

Natural processes are regulated by hormonal systems. Within the animal kingdom endocrine glands secrete hormones that serve this purpose. In mammals, the thyroid, pituitary, and adrenal glands all produce them, as do the pancreas, testes, and ovary. Once secreted, these hormones find their way to receptors, such as the androgen and estrogen receptors, and elicit physiological and biochemical responses (see Box 15.1).

In what follows, the term *endocrine disruption* will refer to "a hormonal imbalance initiated by exposure to a pollutant which leads to alterations in the development, growth, and/or reproduction in an organism or its progeny" (Goodhead and Tyler, in Walker 2009, chap. 15). In recent years there has been a growing concern about pollutants that act as endocrine disruptors. There has also been interest in naturally occurring estrogens and androgens other than those mentioned above, which are found, for example, in surface waters.

The most widely studied of endocrine-disrupting chemicals (EDCs) have been pollutants that act upon the processes of reproduction. Prominent among these have been estrogens that can feminize male fish. A few antiandrogens have also been identified. These have a similar effect to estrogens because they can retard the action of male hormones.

In earlier chapters reference was made to pollutants that can have harmful effects upon reproduction, and it has been suggested that some of these act as disruptors of the hormonal system. In this category is tributyl tin (TBT), which can masculinize female dog whelks (Chapter 11). The DDT metabolite p,p′-DDE provides another example (Chapter 9). The difficulty in both of these cases is some uncertainty about the precise mode of action.

Female dog whelks develop penises when exposed to TBT. It has been suggested that this is due to an increase in the level of the hormone testosterone, resulting from an effect of TBT upon steroid metabolism. If this is true, then TBT can be described as an endocrine disruptor. This issue will be discussed further later in the present chapter. Eggshell thinning in birds caused by p,p′-DDE results from a failure of calcium transport into the shell gland. There is also evidence indicating an effect of p,p′-DDE upon the level of prostaglandin, a chemical messenger that can regulate certain physiological processes. It is not secreted by an endocrine gland. If the primary cause of eggshell thinning is a disturbance in the level of prostaglandin, then this is an example of disruption of a messenger system. If, on the other hand, it is a direct effect upon the transport system for calcium (e.g., the enzyme ATP-ase), it does not answer to this description.

BOX 15.1 ESTROGENS AND ANDROGENS

In vertebrates, estrogens and androgens are hormones that regulate the development of secondary sexual characteristics in females and males, respectively. Estrogens (e.g., estradiol) are secreted by the ovary, and interact with estrogen receptors to promote the development of secondary female characteristics. Androgens (e.g., testosterone) are secreted by the testes, and interact with androgen receptors to promote the development of secondary male characteristics.

Certain pollutants also interact with these receptors, and may do so in two distinct ways. On the one hand, they may mimic the action of the normal hormones, in which case they are termed *agonists*. By contrast, some pollutants interact with receptors and oppose the action of normal hormones, and these are termed *antagonists*.

It follows that a pollutant that acts as an agonist toward estrogen receptors may have a similar effect to another pollutant that acts as an antagonist toward androgen receptors. In both cases the pollutant has a feminizing effect. This phenomenon has been observed in fish exposed to pollutants (Goodhead and Tyler, in Walker 2009, chap. 15).

THE FEMINIZATION OF FISH CAUSED BY ESTROGENS

A critical early study in the UK established that there was a high incidence of intersex in roaches living in wastewater treatment works (see Goodhead and Tyler, in Walker 2009). The term *intersex* describes a condition in which male fish have become feminized. The feminine characteristics include the appearance of ova in the testes. Males so affected are liable to be infertile. Further work established that water from these works was sometimes highly estrogenic (see Purdom et al. 1994). Subsequently, utilizing cell-based bioassays (Table 2.1 and Box 15.2), fractionation of these waters identified three major estrogenic substances. Two of these were the natural steroidal estrogens E2 and E1. The third substance was ethinylestradiol (EE2), a component of the contraceptive pill (Desbrow et al. 1998; Rodgers-Gray et al. 2001). These compounds are all excreted by the human population and are not completely removed by sewage treatments.

EE2 is a highly estrogenic molecule that is moderately persistent in surface waters (see Figure 15.1). Thus, there is concern about the effects that it may have upon male fish in river water below sewage outfalls. Later experiments have provided evidence of the environmental damage that EE2 may cause. In laboratory studies, it has been shown to cause reproductive failure in zebrafish experiencing lifetime exposure to 5 ng EE2/L (Nash et al. 2004). More recently, an entire experimental lake in Canada was treated with 5–6 ng/L of EE2, causing total population collapse of the fathead minnow over a seven-year period (Kidd et al. 2007).

BOX 15.2 VITELLOGENIN

When estrogens interact with estrogen receptors a protein called vitellogenin is released. Vitellogenin is a female yolk protein that can be produced by male fish when they are exposed to estrogens (Purdom et al. 1994). Measuring the release of vitellogenin has proved to be a valuable biomarker assay of exposure both in vivo and in vitro.

This response can be measured in fish that are exposed to polluted waters. The use of cell-based assays (see Chapter 2) aided the fractionation of polluted waters from wastewater treatment works that led to the isolation of EE2 and two natural estrogens (Rodgers-Gray et al. 2001). In such assays cultures of fish cells or yeast cells containing estrogen receptors are exposed to estrogens, and then vitellogenin is released. The quantity of vitellogenin can then be determined as a measure of the strength of this response.

ENDOCRINE DISRUPTION INVOLVING THE Ah RECEPTOR

Earlier in this text the phenomenon of Ah receptor-mediated toxicity was described (Chapter 13). Coplanar polychlorinated biphenyls (PCBs) and dioxins are persistent flat molecules that can bind strongly to the Ah receptor, which is found in the cytosol of vertebrates. When they do so, the degree of binding has been related to the extent of several toxic effects. Endocrine disruption is one of these effects. The degree of such binding can be expressed as dioxin equivalents (Box 13.1).

In a study of the complex pollution of the Great Lakes of Canada and the United States during the latter years of the twentieth century, the breeding success of double-crested cormorants and Caspian terns was found to be inversely related to the dioxin equivalents present in their eggs. Unfortunately, it was not possible at that time to determine to what extent, if at all, endocrine disruption contributed to the breeding failures. That said, there was a clear relationship between the values of dioxin equivalents and the extent of breeding failure, and it seems likely that endocrine disruption contributed to this effect.

Another example of a complex pollution study in North America was the investigation of the pollution of Lake Apopka in Florida (Guillette et al. 1996). After a pesticide spill in 1980, effects upon the endocrine systems of American alligators

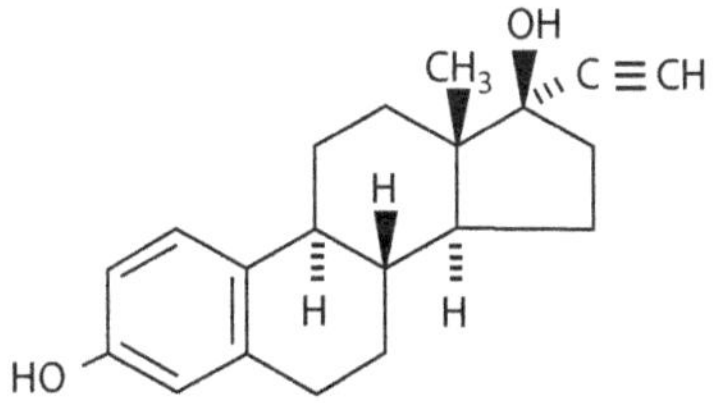

FIGURE 15.1 Ethinylestradiol (EE2) is a potent synthetic estrogen bearing a structural resemblance to estradiol.

inhabiting the lake were reported. Initially these were attributed to effects of DDT metabolites, especially p,p′-DDE, which remained at high levels in the lake over a period of years. Subsequently, attention was focused on relatively high levels of organochlorine insecticides other than DDT, and also of PCBs, in the same ecosystem. In this complicated scenario the interpretation of residue data became controversial. Once again, however, there was evidence to suggest that PCBs might be acting as endocrine disruptors.

DISTURBANCES OF THE METABOLISM OF STEROID HORMONES BY POLLUTANTS

The metabolism of sex steroids such as estradiol and testosterone can be disrupted by pesticides. Forms of the oxidative enzymes called cytochrome P450s are involved in this metabolism (see Box 2.1), and some can be inhibited by EBI fungicides. Examples of this are given by Goodhead and Tyler (2008). Thus, some environmental chemicals have the potential to act as EDCs through disturbance of sex steroid metabolism.

This is a difficult and technical subject that is controversial. It would not be appropriate to describe it in detail in a general account such as the present one, so it will only be dealt with briefly.

One important enzyme of this kind is aromatase (CYP 19), which has a critical role in the conversion of androgens into estrogens. Inhibition or induction of aromatase can change the rate of production, and consequently the cellular levels of steroid hormones. Morcillo et al. (2004) have shown that tributyl tin (see Chapter 11) can inhibit aromatase activity in vertebrates and aquatic invertebrates. It has been suggested that this is the reason why tributyl tin causes imposex in molluscs such as the dog whelk. The inhibition of aromatase may slow the metabolism of testosterone with consequent buildup in the concentration of testosterone in these molluscs, leading to masculinization (Chapter 11). This is the condition of imposex, where female molluscs develop penises and become infertile. The disappearance of dog whelks from stretches of the British coast was attributed to the effects of tributyl tin applied as a descalant to boats.

THE DIVERSITY OF ENDOCRINE-DISRUPTING CHEMICALS (EDCs) IN THE ENVIRONMENT

Members of the plant kingdom produce estrogens that can be found in the aquatic environment. These include the mycoestrogens of fungi and the phytoestrogens of higher plants. Thus, there are naturally occurring compounds that add to the extensive array of EDCs that have been detected in the natural environment. To widen the discussion, some of the more important examples of anthropogenic EDCs will now be briefly reviewed.

Alkylphenol ethoxylates are nonionic surfactants produced by the chemical industry. The breakdown of them leads to the release of alkylphenols such as nonylphenol. Alkylphenols have estrogenic activity. Although this activity is relatively weak, the high concentrations of alkylphenols sometimes found in industrial and domestic effluent are sufficient to cause a feminizing effect upon fish (Sheahan et al. 2002).

Phthalates are said to be among the most abundant synthetic chemicals found in the natural environment. They are used in lubricating oils, the manufacture of plastics, and cosmetics, and are found in substantial amounts in some landfill sites. Residues of them have been found in surface waters and in drinking water. Some phthalates show estrogenic activity. Particularly high levels of phthalates have been found in river water in the developing world (see Goodhead and Tyler 2009). There is evidence that these compounds can have estrogenic effects.

Bisphenol A was first discovered as an estrogen. Subsequently, it came to be used in the production of polycarbonate plastic. Bisphenol has been identified as an environmental pollutant with estrogenic activity.

This catalogue could be extended. Suffice it to say that complex mixtures of diverse EDCs can be found in environmental samples, and questions inevitably arise about possible potentiation of toxicity (see Box 2.2). A similar situation exists with effluents containing mixtures of pharmaceuticals, albeit at low concentrations. Although concentrations of individual pollutants are usually quite low, there is concern that there may be substantial increases in toxicity as a consequence of synergistic interactions. This problem is discussed in Chapter 2, and will be returned to in Chapter 19.

This complicated scenario is rather typical of the pollution problems of today and contrasts with earlier times when large numbers of birds were being lethally poisoned by highly toxic organochlorine or organophosphorous insecticides and dead bodies were conspicuous in agricultural areas, and lethal levels of chemicals—or their effects (e.g., cholinesterase inhibition)—were readily measurable. Nowadays the discovery of pollution problems at an early stage depends on the application of sophisticated technologies by ecologists, chemists, biochemists, and molecular biologists.

SUMMARY

Endocrine disruption has been defined as "a hormonal imbalance initiated by exposure to a pollutant which leads to alterations in development, growth, and/or reproduction in an organism or its progeny." Many of the examples of endocrine disruption in the natural environment have involved pollutants that act as estrogens or androgens. There is evidence to suggest that the development of imposex by dog whelks exposed to TBT may be the consequence of endocrine disruption.

An important example of endocrine disruption has been the feminization of male fish observed in the vicinity of sewage outfalls. This has been attributed to the effect of EE2, a constituent of contraceptive pills. Prolonged exposure to low levels of this estrogen has been shown to cause reproductive failure in zebrafish. In Canada the fathead minnow population of a lake collapsed after seven years of exposure to a low concentration of EE2. This illustrates further an important principle in ecotoxicology: populations may decline or even reach localized extinction due to sublethal effects of pollutants.

Some pollutants that can cause Ah receptor-mediated toxicity act as endocrine disruptors, e.g., dioxins and coplanar PCBs.

Other pollutants that can act as endocrine disruptors include phthalates, bisphenol A, and the alkylphenol ethoxylates, which are formed when detergents break down.

FURTHER READING

Goodhead, R.M., and Tyler, C.R. 2009. Endocrine-disrupting chemicals and their environmental impacts. In *Organic pollutants: An ecotoxicological perspective*, ed. C.H. Walker, chap. 15. 2nd ed. Boca Raton, FL: Taylor & Francis. A wide-ranging review of endocrine-disrupting chemicals in the natural environment.

Kidd, K.A., Blanchfield, P.J., Mollis, K.H., et al. 2007. Collapse of a fish population after exposure to a synthetic estrogen. *Proceedings of the National Academy of the United States of America* 104: 8897–8901. A striking example of breeding failure caused by an EDC.

Nash, J.P., Kime, D.E., van der Ven, L.T.M., et al. 2004. Long-term exposure to environmental concentrations of the pharmaceutical ethynyl estradiol causes reproductive failure in fish. *Environmental Health Perspectives* 112: 1725–1733. Describes effects of long-term exposure of fish to EE2.

Tyler, C.R., Jobling, S., and Sumpter, J.P. 1998. Endocrine disruption in wildlife: A critical review of evidence. *Critical Reviews in Toxicology* 28: 319–361. A useful review of endocrine disruption in wildlife.

16 Anticoagulant Rodenticides

INTRODUCTION

Rats and mice have long been regarded as pests in both rural and urban areas, and over the years a number of different rodenticides have been used to control them. These have included zinc phosphide, which degrades to release the toxic gas phosphine, the natural product strychnine, and the cyclodiene endrin, which is a stereoisomer of dieldrin. However, one group of compounds has come to dominate this market—the anticoagulant rodenticides (ARs). The best known of these is warfarin, which is also used in human medicine. The following account is devoted to them.

As with many other successful pesticides, warfarin was modeled on a natural product. Early in the twentieth century in the United States there was an investigation of the cause of hemorrhaging in cattle. It was discovered that this was being caused by dicoumarol, a compound present in spoiled clover on which cattle sometimes feed. The structure is shown in Figure 16.1 alongside those of warfarin and the superwarfarins brodifacoum and flocoumafen.

MODE OF ACTION OF THE ANTICOAGULANT RODENTICIDES (ARS)

The structures of warfarin and two related superwarfarins are shown in Figure 16.1. In the first place, it can be seen that they bear some *structural resemblance* to dicoumarol, a natural anticoagulant. All possess linked six-membered rings, drawn as hexagons, to which oxygen atoms are attached. On the left-hand side of all four structures are two linked rings, the outermost of which (left-hand side) has an aromatic ring referred to as a benzene ring.

Looking now at the structure of vitamin K at the bottom of the figure, it can be seen that this also has two linked rings, the outermost of which is a benzene ring. All five structures shown in Figure 16.1 have linkages of one kind or another between the ring and oxygen. In warfaring the two superwarfarins and dicoumarol the second of these two linked rings (right-hand side) contains one oxygen atom, in replacement of one of the carbons. This is described as a heterocyclic ring because not all of the atoms forming it are carbon atoms.

In the livers of vertebrates vitamin K binds to a site on an enzyme termed carboxylase, which is involved in the synthesis of clotting proteins. Because of their structural similarities to vitamin K, dicoumarol, warfarin, and the superwarfarins can also

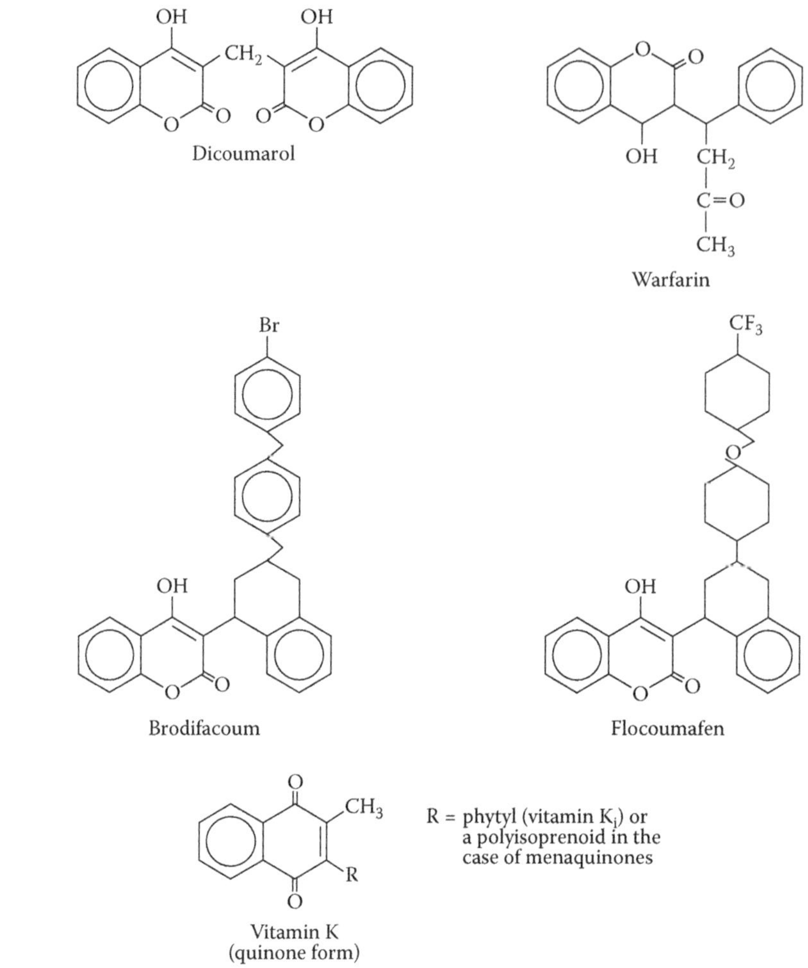

FIGURE 16.1 Anticoagulant rodenticides.

bind to this site—and when they do so, they can exclude vitamin K. They can act as antagonists and block the formation of fully synthesized clotting proteins.

These events are presented in a simplified way in Figure 16.2. The carboxylase enzyme, for which vitamin K is a cofactor, exists in the liver and has the role of completing the synthesis of clotting proteins. Under normal conditions, vitamin K is converted into its reduced state and then the carboxylase enzyme interacts with the precursor of a clotting protein. This interaction leads to the completion of the synthesis of the clotting protein, which can then be exported from the liver into the blood. This interaction involves the conversion of the vitamin K back into the reduced state. After this, the cycle is repeated to complete the synthesis of another molecule of the clotting protein.

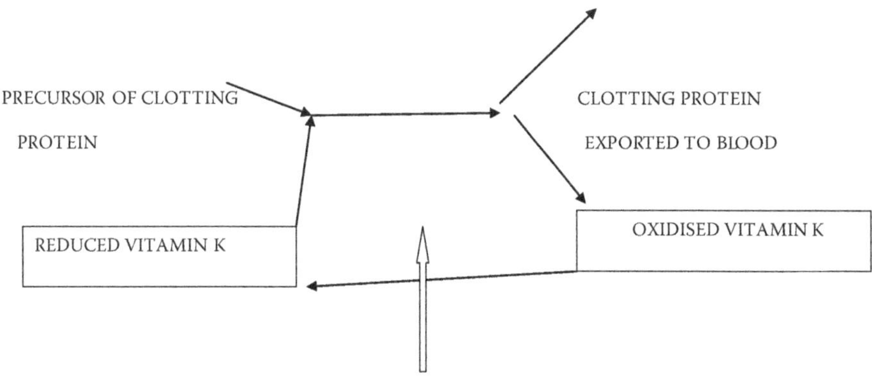

FIGURE 16.2 Mode of action of warfarin.

When ARs attach themselves to the binding site for vitamin K, this essential cofactor does not return to the oxidized state, and carboxylation of the precursor of the clotting protein stops. An uncarboxylated precursor of the clotting protein is released into the blood. Initially this interaction leads to no harmful effect upon animals, because there is still plenty of normal functional clotting protein in the blood. However, these ARs attach themselves very strongly and persistently to the vitamin K binding site, and with time, as the level of the normal clotting protein in blood slowly falls, it is replaced by a dysfunctional uncarboxylated clotting protein. In rats and mice receiving a lethal dose of a rodenticide it is typically five days or more before symptoms of poisoning show. Death is eventually caused by hemorrhaging. When injuries occur to blood vessels and clotting fails there is uncontrolled bleeding.

In theory there is a very suitable biomarker assay for this type of poisoning—an antibody to measure the concentration of the unfinished clotting protein in blood of protected species (e.g., owls) that may be exposed to ARs. To the best of the knowledge of this author, no such bioassay is currently available. The environmental side effects of ARs will be discussed in later sections.

SECONDARY POISONING OF PREDATORS AND SCAVENGERS

Warfarin was once widely used in Western countries to control rats and mice on farms and on domestic and industrial premises in urban areas. During the latter part of the twentieth century warfarin-resistant rodent strains began to emerge and other ARs were introduced to control them. Second-generation ARs, sometimes called superwarfarins, have included difenacoum, bromodiolone, brodifacoum, and flocoumafen. Some structures are shown in Figure 16.1. These act by the same mechanism as warfarin, but bind more tenaciously to sites in the liver (including the site that binds vitamin K) than does warfarin itself. They tend to be more persistent and toxic than warfarin, although there is considerable variation in toxicity between species. The acute oral median lethal dose (LD_{50}) for warfarin in the rat has been estimated as

1 mg/kg, whereas flocoumafen and brodifacoum show values between 0.25 and 27 in mammals and between 0.25 and >300 in birds. Some superwarfarins show half-lives of 100 days or more in the livers of rodents or birds during the later stages of their elimination (see Huckle et al. 1989; Fergusson 1994).

Because of their tendency to persist in the liver, there has been concern about their effects upon predators and scavengers that feed upon rodents. Included here are owls, buzzards, red kites, carrion crows, magpies, jays, and other members of the crow family (corvids). Mammalian predators such as stoats, weasels, and polecats also appear to be vulnerable.

A critical question is: What levels of these superwarfarins in the livers of predators and scavengers are indicative of poisoning? Newton et al. (1990) fed mice containing brodifacoum to barn owls that died within six–seventeen days of commencement of dosing. These birds consumed between 0.150 and 0.182 mg/kg of the rodenticide, and were found to contain 0.63–1.25 mg/kg in the liver. In a field trial with brodifacoum conducted in Virginia, screech owls were exposed to brodifacoum (Hegdal and Colvin 1988). Birds found dead five–thirty-seven days posttreatment contained 0.4–0.8 mg/kg in the liver—very similar to the levels found in the laboratory study with barn owls. Taken together, owls poisoned by brodifacoum in these studies had residues of the rodenticide in liver falling within the range of 0.4–1.25 mg/kg.

Concern has been expressed about the widespread use of superwarfarins to control rodents on farms in Western countries. In the UK, however, a widespread decline in the barn owl could not be attributed to the use of superwarfarins (Newton et al. 1990). Only a small proportion (2%) of barn owls found dead during the period in question (1983–1989) contained residues of brodifacoum plus diphenacoum of 0.1 ppm or more. At that time the use of these and other superwarfarins was restricted to the interior of buildings in order to minimize hazards to owls and other predators or carrion feeders that consume rodents.

Anticoagulant rodenticides have been used in some conservation areas to control rodents and vertebrate predators, sometimes with unwanted side effects. In New Zealand, the use of brodifacoum to control rodents and vertebrate predators on conserved islands has led to problems (Eason and Spurr 1995; Eason et al. 2002). Casualties included native raptors such as the Australasian harrier and other species, such as the pukeko and western weka. In one case an entire island population of western weka was wiped out following the use of brodifacoum. Both primary poisoning due to consumption of bait and secondary poisoning due to feeding upon poisoned prey/carrion were reported.

In Canada a similar case was reported (Howald et al. 1999). On Langara Island, British Columbia, brodifacoum was used to control rats. Ravens were lethally poisoned after consumption of bait containing the rodenticide. They were also affected by secondary poisoning due to the consumption of poisoned rats. Both live and dead rats were eaten, with ravens being predators as well as carrion feeders. These incidents illustrate the dangers of misusing such a highly poisonous rodenticide in conservation areas.

RESISTANCE TO ANTICOAGULANT RODENTICIDES

Resistance to warfarin in Norway rats was evident within two decades of the pesticide's registration. Subsequently, resistance was also reported in black rats and house mice. As with many other pesticides, repeated use led to the emergence of resistant strains—as the consequence of what might be termed unnatural selection. This resistance has not necessarily been shown to other rodenticides. For example, rats showing resistance to warfarin have often retained their susceptibility to diphenacoum and bromodiolone (Shore and Rattner 2001). So far as this author is aware, resistance has not yet been reported to brodifacoum.

Much of the known resistance to warfarin and certain other anticoagulant rodenticides has been attributed to mutant forms of the enzyme that binds vitamin K in the liver of rodents (see Figure 16.2 and earlier discussion). In some cases a mutant form found in resistant rats binds warfarin far less strongly than does the form found in normal susceptible rats. In consequence, warfarin is not very effective in inhibiting the production of clotting proteins in such resistant strains. In another example, the mutant form binds warfarin strongly—but the binding is not sustained—so again warfarin is rendered ineffective (Thijssen 1995). In principle, these examples are similar to many cases of resistance to insecticides by insects—where resistance is due to the existence of mutant forms of the target, e.g., acetylcholinesterase, sodium channel, GABA receptor, etc. These examples are discussed in earlier chapters.

By contrast, examples are also known of anticoagulant resistance associated with enhanced metabolic detoxication in the resistant strain (see Walker 2009, sect. 11.6.2). Once again, there is evidence for the involvement of forms of cytochrome P450 in resistant rats (Markussen et al. 2008). These examples are comparable to many cases of insecticide resistance in insects that are due to enhanced detoxication—as discussed in earlier chapters.

SUMMARY

Most rodenticides are anticoagulants and related to the widely used compound warfarin. Structurally, they are related to a naturally occurring anticoagulant dicoumarol, which occurs in clovers and some grasses.

These compounds act as vitamin K antagonists in the liver. They compete with vitamin K for its binding site on a carboxylase enzyme. By doing so, they prevent the completion of synthesis of clotting protein, and the liver ceases delivering this vital biochemical to the blood. The consequences of this are not immediately apparent because it takes some time for the levels of normal clotting protein to fall in the blood.

After a period of days, however, the levels of normal functional clotting protein become so low that the blood is no longer able to clot. When blood vessels are damaged, there is uncontrolled bleeding. Animals eventually die from hemorrhaging.

Over time, strains of rat have emerged that are resistant to warfarin, and a new generation of rodenticides related to this compound has emerged. Some of these, such as brodifacoum and flocoumafen, are highly toxic and persistent, and their use has been restricted because of the perceived hazards that they present to wild predators and scavengers. Brodifacoum, for example, has been used to control rats and vertebrate predators in conserved areas of New Zealand and Canada, sometimes with undesired side effects.

FURTHER READING

Eason, C.T., Murphy, E.C., Wright, G.R.G., et al. 2002. Assessment of risks of brodifacoum to non target birds and mammals in New Zealand. *Ecotoxicology* 11: 35–48. Describes unwanted side effects of a new generation anticoagulant rodenticide used in conservation areas.

Hegdal, P.L., and Colvin, B.A. 1988. Potential hazard to eastern screech owls and other raptors of brodifacoum bait used for vole control in orchards. *Environmental Toxicology and Chemistry* 7: 245–260. Evidence for secondary poisoning of owls caused by brodifacoum.

Markussen, M.D., Heiberg, A.C., Fredholm, M., et al. 2008. Differential expression of cytochrome P 450 genes between bromodiolone-resistant and anticoagulant susceptible Norway rats. A possible role for pharmacokinetics. *Pest Management Science* 64: 239–248. Evidence for a resistance mechanism to bromodiolone in Norway rats.

Newton, I., Wyllie, I., and Freestone, P. 1990. Rodenticides in British barn owls. *Environmental Pollution* 68: 101–117. A study of rodenticide residues in barn owls found dead in the field.

Shore, R.F., and Rattner, B.A., eds. 2001. *Ecotoxicology of wild mammals.* Chichester, UK: John Wiley. A substantial reference work describing the effects of various pesticides, including anticoagulant rodenticides, on wild mammals.

Thijssen, H.H.W. 1995. Warfarin-based rodenticides: Mode of action and mechanism of resistance. *Pesticide Science* 43: 73–78. A valuable account of mode of action of anticoagulant rodenticides.

Section III

Further Issues

Section I of the foregoing text focussed on basic principles of ecotoxicology which were then illustrated by examples in Section II. This final section addresses wider issues. Chapter 17 considers pollution problems from a global perspective. Chapter 18 discusses the question of legislation to control environmental pollution. Chapter 19 describes some current pollution problems and attempts to look into the future.

17 Pollution Problems Worldwide

INTRODUCTION

When the island of Krakatoa in Indonesia was blown apart by a huge volcanic explosion in 1883, the consequences were evident over a large area of the globe. The explosion was heard in places far away. Locally there were disastrous tsunamis causing the loss of many lives. Volcanic ash was blown to a height of c. 80 km and encircled the world for a long time afterwards in the stratosphere. It has been estimated that the global temperature fell by c. 1.2°C, and did not return to normal until 1888. This was a consequence of the presence of volcanic ash circulating in the stratosphere and restricting the entry of solar energy into the atmosphere. This cataclysmic event led to pollution not just locally, but also on a global scale, which persisted for a lengthy period.

Thus far the present text has focused on basic principles and issues of ecotoxicology. These have been illustrated by taking examples of particular types of pollutants describing environmental problems that they have caused. In this chapter some wider issues will be addressed, some of which have been touched upon in earlier chapters. Phenomena like global warming caused by greenhouse gases and thinning of the ozone layer are matters of international concern that need to be resolved through international cooperation (Chapter 7). This also applies to the problem of acid rain, which is discussed in the same chapter. Sulfur dioxide has been a major contributor to acid rain in the northern hemisphere, and questions have been asked about its role in damaging forests and acidifying lakes (Howells 1995). North America and Scandinavia have been particularly affected, and there has been political discussion about the extent to which the export of sulfur dioxide from one country to another has been responsible. With time, emissions of sulfur dioxide and other gases contributing to acidification of rain have been reduced—notably since the fall of communism. The Gaia theory is a concept that attempts to look at natural processes and cycles in a holistic way, considering the earth to be a single self-regulating entity. Hopefully the use of models of this kind will help to resolve global pollution problems.

Much of the foregoing text has been devoted to man-made (anthropomorphic) pollutants. Like the natural pollutants released by volcanoes, these do not respect international boundaries. The movement of pollutants across the boundaries between states has caused political problems. Examples include the "export" of acid rain to Scandinavia, the wide dissemination of radionuclides following disasters at nuclear power stations, and pollutants released into international waters by boats. The present chapter addresses these and other wider issues that were not dealt with in the more focused approach of the two earlier parts of the book.

INTERNATIONAL APPROACHES TO POLLUTION PROBLEMS

Some international organizations have had an important role in addressing pollution problems that transcend national boundaries. The World Health Organization, for example, has been involved in the large-scale use of pesticides to control vectors of disease, such as the malarial mosquito and the tsetse fly. The Food and Agriculture Organization of the United Nations has played a role in promoting the safe use of pesticides to control pests, diseases, and weeds in agricultural areas of developing countries. The Intergovernmental Panel on Climate Change (IPCC) has been involved in monitoring global warming. The International Atomic Energy Agency (IAEA) has investigated radiation hazards and the misuse of chemicals in developing countries. Scientific bodies such as the Scientific Committee on Problems of the Environment (SCOPE) and the Society of Environmental Toxicology and Chemistry (SETAC) have been involved in the organization of international meetings and publishing proceedings and journals that address pollution problems worldwide.

DEVELOPED AND DEVELOPING COUNTRIES

Until now, much has been said about the environmental side effects of pesticides and other man-made chemicals that have been discovered, used, and studied in the developed world, and many of the examples have been of pollution problems in Western Europe and North America. Much less has been said about pollution problems in Africa, India, and South America because little has been published on this subject in leading international journals.

Globally, the use of pesticides was very limited before the World War II. After 1945, however, the pesticide industry grew rapidly in Western Europe and North America, and in time, problems began to surface. Resistant strains of insect pests began to emerge with the overuse of insecticides. New pest species (e.g., red spider mites) started to gain ground where natural enemies were depleted due to excessive application of insecticides. Also, resistant strains of insect pests gained ground due to overuse of these chemicals. By the early 1960s environmental problems were encountered with persistent pesticides such as certain organochlorine insecticides and methylmercury fungicides (Chapters 9 and 11). These chemicals undergo biomagnification with movement along food chains, leading to harmful effects upon predators belonging to the highest levels of these chains.

The recognition of these problems led to the initiation of stricter controls over the marketing of pesticides and other biocides in the developed world. There was concern about protecting the natural environment and maintaining biodiversity. The discipline of ecotoxicology came to be recognized and new testing methods validated to establish the environmental safety of these chemicals. Chapter 18 is devoted to the question of the testing methods used to establish the environmental safety of pesticides.

This trend was, however, largely restricted to the developed world. In lands visited by famine and epidemics of deadly diseases such as malaria or sleeping sickness, there was not the same concern about protecting the natural environment. Protecting ecosystems against harmful effects of pesticides was less important than saving human lives. Thus, there were far fewer impediments to using hazardous pesticides than in Western Europe during the latter part of the twentieth century. There were many instances of serious damage to ecosystems over large areas during pest control or vector control programs, but much of this was not well recorded.

Many of the pesticides banned in developed countries continued to be used in developing countries because they were effective and affordable. Among the most striking examples of this were aerial applications of highly toxic insecticides for control of vectors of disease and locusts in Africa (hazardous to spray operators as well as to the natural environment). These will be discussed in the next section. Such compounds were also widely used for control of agricultural pests in more conventional ways by operators on land.

Mullie et al. (1991) reported on the large-scale use of pesticides in Senegal during the late 1980s on rice and sugar cane. Included among these were dieldrin and carbofuran, two compounds regarded as environmentally hazardous. Dieldrin, of course, was subject to extensive bans in developed countries by that time. They express concern about careless disposal of unused pesticide. There was evidence for low population densities of birds, notably waders, in areas treated with pesticides.

Douthwaite (1992) reported that despite restrictions on the agricultural use of DDT in Zimbabwe in 1985, some use of it continued into the late 1980s. A survey of Lake Kariba during the period 1989–1990 showed that there were substantial residues of DDT and its metabolites in the eggs of fish eagle (*Haliaeetus vocifer*). There was also evidence of a substantial degree of eggshell thinning in these eggs (<20%). For further discussion, see Chapter 9.

An investigation into the continuing use of persistent organochlorine insecticides in Africa was carried out in a joint program run by FAO and IAEA in the early 1990s (see IAEA 1997). In this publication it is stated that compounds such as DDT, heptachlor, dieldrin, chlordane, and toxaphene were still widely used at that time, in ways that they were not in the developed world.

The problem was twofold. On the one hand, there were fewer restrictions on these practices than in the developed world. Additionally, even where there was legislation, it was not necessarily rigorously implemented. There were, evidently, considerable differences in practice between individual African states. Some African states, e.g., Kenya, Tanzania, South Africa, and Uganda, have substantial incomes from ecotourism. For them, damage to wildlife caused by misuse of pesticides is a matter of economic importance.

AERIAL APPLICATION OF PESTICIDES

Aerial application of pesticides is a very effective way of treating large areas of land in a short time. It has, however, the disadvantage that it is very difficult to apply the pesticide accurately to the treated area. Spray drift can be a serious problem. Even light winds can cause spray to land outside of the designated area—with consequent risk to humans as well as the natural environment when dealing with hazardous pesticides (e.g., chlorinated cyclodiene insecticides such as dieldrin or endrin). Pilots have died as a consequence of flying through their own spray clouds.

BOX 17.1 SOME INSECTS CONTROLLED BY LARGE-SCALE SPRAYING

Large-scale spraying of insecticides, often from airplane or helicopter, has been carried out in developing countries to control vectors of disease and agricultural pests. These operations can cause widespread environmental damage, including the poisoning of wild mammals and birds. Compounds no longer marketed in developed countries have frequently been used, including dieldrin, DDT, heptachlor, and endrin. In recent times it has been difficult to establish to what extent these practices are still followed.

Several different species of tsetse fly (*Glossina* spp.) are found in Africa. They act as vectors for trypanosomes, which cause sleeping sickness (trypanosomaniasis). Koeman et al. (1978) describe the consequences of aerial spraying in Nigeria to control tsetse and give evidence that less environmental damage is caused by manual spraying performed by operators at ground level than by aerial spraying.

Another important vector that has been controlled in this way is the malarial mosquito (*Anopheles* spp.). Spraying has been carried out in areas of Africa, India, and Latin America affected by malaria (Brooks 1974). The mosquitoes breed in lakes and swampy areas and can be controlled by aerial spraying of DDT, that is, until resistant strains of mosquito begin to emerge.

Some species of locust have caused great damage to crops over large areas of land in Africa, Asia, and Australia. Important species include the migratory locust (*Locusta migratoria*), found throughout Africa and South Eurasia, Australia, and New Zealand, and the desert locust (*Schistocerca gregaria*), found from North Africa to North India. Locusts are large grasshoppers that can form huge swarms which cause havoc in agricultural areas. Large swarms can contain many billions of individuals and cover areas of hundreds of square kilometers. Aerial spraying (e.g., with DDT or dieldrin) has offered a rapid and effective method of controlling locust swarms, but has also caused serious and widespread pollution.

Koeman et al. (1978) report on the control of tsetse flies in Nigeria by aerial spraying (Box 17.1). Observations were made during the period 1974–1976. After application of dieldrin by helicopter to one area in 1974, vertebrate animals and birds were found dead containing residues of dieldrin in the liver ranging from 5.0 to 66 ppm—levels associated with acute lethal poisoning. Subsequent surveys showed that some species of birds (e.g., flycatcher species) disappeared or were much reduced in this area. Many invertebrates were found dead after these spray operations. Evidently insectivorous birds and mammals were poisoned by the high levels of dieldrin encountered on insects. There was some evidence that ground spray applications caused less ecological damage than helicopter applications when using insecticides at comparable dose rates. The authors report that dieldrin was widely used in Africa to control tsetse fly at that time. Other organochlorine insecticides in use for tsetse control were DDT and endosulfan.

Apart from these African programs to control locusts and vectors of disease, there have been others directed at vertebrate pests. Some organophosphorous insecticides have been used to control the avian pest *Quelea* (Bruggers and Elliott 1989). Both parathion and fenthion have been applied aerially to roosts in order to control this bird. It has been suggested that this practice may represent a human health hazard as well as an environmental problem, because *Quelea* is eaten by local people.

The environmental problems caused by large-scale aerial spray programs have not been entirely restricted to developing countries. Examples include the aerial application of organophosphorous insecticides (OPs) in North America to wooded areas. The environmental damage caused by spraying almond orchards in California and large tracts of forest in New Brunswick, Canada, have been discussed in Chapter 10.

PLASTICS

Plastics are synthetic polymers that came into large-scale use during the latter part of the twentieth century. Unlike natural polymers like proteins, carbohydrates, or soil organic matter, they are not readily broken down by natural processes, e.g., enzymic action, and are often highly persistent in the environment. Plastics such as polythene and polystyrene are widely used for packaging, and careless disposal of them has led to widespread pollution of the earth's surface (see *Encyclopedia Britannica* 2013).

People who live in coastal areas throughout the world will know that small pieces of plastic are prolific along seashores in many areas of the globe, and it has been reported that such materials have been found in many remote locations, from the bottom of the ocean to the top of Mount Everest. The oceans of the world act as sinks into which small plastic fragments are transported by rivers or by the movement of air.

The environmental hazards presented by this type of pollution are particularly evident in marine habitats. Plastic pollution can be harmful to marine organisms because of entanglement or ingestion. Ingestion of plastic material can be fatal to seabirds, large cetaceans, and marine turtles. An additional problem is that small pieces of plastic can readily adsorb persistent marine pollutants such as organochlorine insecticide residues, polychlorinated biphenyls (PCBs), and dioxins. In this way they can act as vectors, transporting pollutants from the aqueous environment into the bodies of marine animals.

This is a problem that can be addressed in two ways: (1) by greatly improving waste disposal practices and (2) by replacing existing plastics with biodegradable ones, such as are now available to aid waste disposal in some countries.

NANOPARTICLES

Recently interest has grown in the possible toxic effects of small engineered particles (Scown et al. 2010). Nanoparticles have been defined as particles that have one dimension that is less than 100 × 10^{-9} mm. In fact, small colloidal particles fall into the same size range as nanoparticles. Engineered nanoparticles (ENPs) are present in many industrial products, including certain paints, sprays, surface coatings, cosmetics, pesticides, and products for bioremediation of soils. The materials constituting ENPs are diverse. Included here are titanium oxide, aluminum oxide, silver, cerium oxide, zinc oxide, and carbon. Nanoparticles of one kind or another are widely distributed across the global environment.

Nanoparticles occur naturally, and some studies have shown that ENPs account for only a small percentage of this material in nature. For example, Oberdorster (2001) found a large quantity of nanoparticles in air samples, but only a small percentage of this could be described as engineered. Soil clay minerals and soil organic matter (humus) are of colloidal size and contribute to the nanoparticles found in air. This contribution can be large under hot and dry conditions.

There are two distinct ways in which nanoparticles may express ecotoxicity. One is by the toxic effects of the particles themselves. Thc other is by acting as vectors for persistent pollutants, an issue that has already been raised in regard to pollution by small pieces of plastic in the previous section. This subject is still in its infancy, and there are many unanswered questions about the significance of global pollution by ENPs.

SUMMARY

Some pollution problems are global in scale, do not respect international boundaries, and raise issues for the international community. Among these are global warming, acid rain, pollution by radioisotopes following nuclear accidents, pollution of the sea following the wreckage of oil tankers, and widespread dissemination of plastics and engineered nanoparticles. There are international organizations that attempt to address such issues by investigating them, disseminating information about them, and promoting international agreements for pollution control.

Global models such as the Gaia theory attempt to approach these issues in a holistic way. It is to be hoped that they will guide the path to rational solutions at the international level.

Many of the pollution problems described in this book involve industrial chemicals such as pesticides. These have been studied and, to a considerable extent, pollution problems resolved in the developed world. Far less is known about the situation in the developing world, where pesticides have sometimes been used on a large scale to control vectors of disease and major pests of crops. There is less stringent control of pesticides in developing countries than there has been in the developed world. One concern has been the practice of large-scale aerial spraying to control agricultural pests and vectors of disease in developing areas.

FURTHER READING

Brooks, G.T. 1974. *Chlorinated insecticides: Biological and environmental aspects*. Vol. 2. Cleveland, OH: CRC Press. A standard text that gives information on the use of organochlorine insecticides in the developing world.

International Atomic Energy Agency (IAEA). 1997. *Organochlorine insecticides in African ecosystems*. Report IAEA-TECDOC-931. IAEA and FAO. Reviews the position regarding the use of these persistent pesticides in Africa in the 1990s.

Koeman, J.H., Den Boer, W.M.J., Feith, H.H., et al. 1978. Three years observation on side effects of helicopter applications of insecticides used to exterminate *Glossina* species in Nigeria. *Environmental Pollution* 15: 31–59. A detailed paper on the aerial application of insecticides in Nigeria during the 1970s.

Scown, T.M., Van Aele, R., and Tyler, C.R. 2010. Do engineered nanoparticles present a threat in the aquatic environment? *Critical Reviews in Toxicology* 10: 653–670. A concise review on nanoparticles in the aquatic environment.

18 Risk Assessment and Legislation

INTRODUCTION

Ecotoxicity tests that have been used during the course of statutory risk assessment of environmental chemicals were described in Chapter 2. In many of them the endpoint is lethality, in order to obtain values for median lethal dose (LD_{50}), median lethal concentration (LC_{50}), etc. Tests have also been employed that work to nonlethal endpoints. Ecotoxicity data of this kind can be employed to estimate risk.

In the ensuing description the following definitions will be used: (1) *hazard* is the potential to cause harm, and (2) *risk* is the probability that harm will be caused. Sometimes the literature is inconsistent in using these terms and, *risk assessment* and *hazard assessment* are used synonymously.

RISK ASSESSMENT OF CHEMICALS

In developed countries, pesticides and other biocides destined to be released into the environment are subject to more rigorous testing methods than are industrial chemicals in general, before they can be marketed or used. The following discussion will be concerned primarily with the use of ecotoxicity data for the purposes of the *statutory* risk assessment of pesticides and other environmental chemicals, i.e., risk assessment that is required by a legislative body.

In ecotoxicology the central interest is in establishing the levels of environmental chemicals that pose a risk at the *level of population* (see Walker et al. 2012, chap. 17). Unfortunately, for reasons of cost, such full-scale environmental risk assessment cannot be undertaken as part of the normal statutory testing of chemicals, although it may be carried out in certain cases of particular concern to regulatory authorities. For example, field trials of a new pesticide may be required if there are uncertainties arising from normal risk assessment (see Somerville and Walker 1990).

Typically, in statutory risk assessment, a comparison is made between two things:

1. The toxicity of a chemical to an organism, expressed as a concentration, e.g., an EC_{50} or NOEC. (An EC_{50} is the concentration of a chemical that will have a designated effect (e.g. on an enzyme activity or on behaviour) that is shown by 50% of the animals tested. An NOEC is a 'no observable effect on concentration.)
2. The anticipated exposure of an organism to the same chemical expressed in the same units as are used for the ecotoxicity data, i.e., concentration in water, food, or soil to which the organism is exposed.

Using such data a ratio may be calculated, for example:

Predicted no effect concentration (PNEC)

The ratio of PEC/PNEC is a risk quotient. If the value is well below 1, the risk is regarded as being low; if it is 1 or more, the risk is regarded as being high.

The limitations of this approach are not difficult to see (see Chapter 2). It is seldom possible to perform tests upon the species deemed to be most at risk in the natural environment, so surrogate species are used for testing. There are often considerable differences between species in their susceptibility to poisoning by any one chemical. Consequently, the value of results obtained using surrogate species in risk assessment is subject to great uncertainty. Also, the developmental stage that is most susceptible to the toxic action of a chemical may not be the one that is used for testing; e.g., animals can be most susceptible to a chemical when they are at a juvenile stage, but this is not usually included in a testing protocol for reasons of feasibility or cost. A further problem may be the actual endpoint used in a test. A lethal toxicity test may give no indication of sublethal effects that could be of critical importance to the survival of an organism in the field, as is explained in earlier chapters. This can be very important for neurotoxic chemicals. Many insecticides (e.g., organochlorine, organophosphorous, pyrethroid, and neonicotinoid insecticides) are neurotoxic. Because of uncertainty about the value of ecotoxicity data, large safety factors need to be incorporated when estimating risk. Sometimes an EC_{50}, for example, may be divided by 1000 to obtain a PNEC value. Another problem is the uncertainty about the environmental exposure to a chemical that will actually occur if it is used as a pesticide. A PEC value may be rather different from the environmental concentration reached when a pesticide is actually applied in the field.

If the statutory risk assessment of a novel pesticide is inconclusive, then it may be necessary to carry out further work before a definite decision can be made. This may need to take the form of a field trial that will give evidence of the actual effects of that chemical under realistic conditions (see Chapter 4). However, as mentioned before, such trials are seldom undertaken because of considerations of cost.

STATUTORY REQUIREMENTS FOR RISK ASSESSMENT OF ENVIRONMENTAL CHEMICALS

There are differences between countries in the legislative requirements for the marketing and use of pesticides and other environmental chemicals. As already mentioned, testing requirements are generally more stringent in developed countries than they are in developing ones. The requirements of the United States, Canada, European Union, and Japan are broadly similar, but there are differences in detail, e.g., the preferred species to be used for toxicity testing. The requirements of the European Union are summarized in Walker (2006).

The situation is different in developing countries. Here, legislation of this kind is very limited, and where it does exist, it may not be rigorously enforced. Pesticides banned in Western countries are sometimes still used in some countries.

In developed countries mammalian toxicity data to establish human health risks are generated as a requirement for the registration of new pesticides. These data may also be used for statutory environmental risk assessment. Data are also usually required for birds and fish. This typically includes the determination of acute oral toxicity, long-term toxicity, and reproductive toxicity for birds, and acute toxicity, long-term toxicity, and effects upon reproduction for fish. One or two species of birds may be used—typically the Japanese or the bobwhite quail and the mallard duck. Tests are also specified for terrestrial invertebrates such as honeybees, earthworms, and beneficial organisms (e.g., predators or parasites that control pests) (Walker 2009).

There is a marked difference between the protocols for testing of pesticides for risks to human health and those followed in ecotoxicity testing. Whereas in human risk assessment several surrogate species are used to assess risk to a single species, in statutory environmental risk assessment a small number of laboratory species are used to assess risks to a large number of wild species; e.g., two species of birds are used to represent hundreds of free-living species. In the latter case, the surrogates are often not closely related to the free-living species deemed to be at risk (see also Chapter 2).

THE REACH PROPOSALS OF THE EUROPEAN UNION

The Registration, Evaluation, Authorisation and Restriction of Chemicals (REACH) proposals of the European Union were implemented in 2006. Prior to that they were the subject of much debate, and there was critical comment in reports of the UK Royal Commission on Environmental Pollution and the Fund for the Replacement of Animals in Medical Experiments (see Combes et al. 2003; Walker 2006). This legislation is particularly concerned with health risks presented by chemicals to humans, but also deals with risks to the natural environment.

In REACH, the tests required for establishing the risks to the natural environment are determined by a tiered system that defines different categories of toxicity testing. This system focuses upon the expected annual level of production or importation of a chemical. The higher the level of production/importation, the more stringent the requirements for ecotoxicity testing. This approach is unsatisfactory from a scientific point of view because the risks posed to free-living organisms depend on levels of exposure, not levels of production. Rates of production are not closely related to levels of exposure for a number of reasons. Above all, the concentration that actually ends up in surface waters or in soils depends upon properties of the chemical itself, including volatility and chemical stability. A tiered system such as this should be based on estimated levels of exposure, not on rates of production or importation. The current system has been preferred to the more scientific one on the grounds of convenience and cost. Here the discussion enters the political arena, and there continues to be lively debate about the acceptable cost of legislation to protect the environment.

That said, there are other issues here. A relatively high level of production of a chemical does not necessarily mean that there will be dangerously high levels in surface waters or soils. If a chemical is unstable either chemically or biochemically, or if it is highly volatile, it is very unlikely to reach high environmental concentrations in locations distant from its point of release. Thus, the REACH system may require unnecessary testing of a chemical, and consequent extra cost for the manufacturers. The escalation of costs of this kind may discourage the production of new, more environmentally friendly pesticides or biocides. It may also raise an ethical concern that unnecessary extra testing may cause suffering to laboratory animals. The question of cost will be discussed further in Chapter 19.

ETHICAL ISSUES

In recent years there has been growing opposition to toxicity testing that causes suffering to laboratory animals (see Chapter 2). A major aim has been to avoid tests that work to the endpoint of lethality. Sometimes national regulations have required the determination of median lethal doses or concentrations (e.g., LD_{50} or LC_{50}) where these are unnecessary. This does not make either scientific or economic sense. Sometimes all that is required is the determination of a dose that does not have a lethal effect. This can be accomplished by following the *fixed-dose procedure*, which has been put forward by the British Toxicology Society (see Timbrell 1995).

SUMMARY

In most countries there are statutory requirements for the risk assessment of chemicals that are to be released into the environment. The performance of ecotoxicity tests is part of this process of risk assessment.

These statutory requirements are more rigorous, and more strictly enforced, in developed countries than they are in most developing ones. In general, there are stricter requirements for the testing of pesticides and biocides than for most other industrial chemicals. This is not surprising because pesticides and biocides are intended to have toxic effects when they are released into the environment. The REACH proposals that have been adopted by the European Union are taken as an example of a legislative system that is being used to regulate the release of chemicals into the environment.

There is an ongoing debate about the improvement of testing methods for environmental risk assessment. Ecotoxicologists have argued for alternative testing procedures that are more rigorous scientifically. Animal welfare organizations have campaigned for the adoption of procedures that will cause less suffering to laboratory animals. These issues will be discussed further in Chapter 19.

FURTHER READING

Combes, R.B., Dandrea, J., and Balls, M. 2003. Registration, Evaluation, Authorisation and Restriction of Chemicals (REACH) proposal FRAME and the Royal Commission on Environmental Pollution common recommendations for assessing risks posed by chemicals under the EC REACH system. *ATLA (Alternatives to Laboratory Animals)* 31: 529–535. A critical overview of the REACH proposals.

Walker, C.H. 2006. Ecotoxicity testing of chemicals with particular reference to pesticides. *Pest Management Science* 62: 571–583. A review of ecotoxicity testing procedures.

19 Current Issues and Future Prospects

INTRODUCTION

During the latter part of the twentieth century a number of pollution problems came to be recognized. In the developed world this discovery led to the introduction of restrictions and outright bans on the marketing and use of certain chemicals. Prominent examples included persistent organochlorine insecticides such as DDT, aldrin, dieldrin, and heptachlor, methylmercury fungicides, and polychlorinated biphenyl (PCB) mixtures. In many cases these measures, in the longer term, led to the decline of environmental residues and some recovery of populations that had been adversely affected. Evidence of this has come from long-term studies in North America and Western Europe, which are described in Section 2 of this book. Thus, bans on DDT were followed by some recovery of affected populations of predatory birds in the area of the Great Lakes of North America, and bans of aldrin and dieldrin in Great Britain were followed by recovery of raptors such as peregrines and sparrowhawks (Chapter 9). Not all has been easy sailing, though. In some badly affected areas of North America unacceptably high levels of pollution have continued to exist even into the present millennium. Examples include continuing pollution by methylmercury, PCBs, and dioxins. Also, there is still concern about the misuse of pesticides in the developing world.

With the success of statutory restrictions—and of more careful use of pesticides—there is the danger that people will assume that serious pollution problems like these can be consigned to the past, in the Western world, at least. Unfortunately, this would be a little overoptimistic. On the one hand, there are still a few long-standing problems with highly persistent pollutants in some locations, which are described in the second part of this book. Residues of persistent polychlorinated compounds (PBCs, dioxins) and of methylmercury are cases in point. Also, in the light of more recent research, other less obvious issues have emerged (e.g., effects of mixtures of environmental chemicals, sublethal effects of neurotoxic compounds on behavior, and the problem of chemicals that cause endocrine disruption).

In this concluding chapter an account will be given of specific pollution problems that have emerged relatively recently and also some topical issues—assessing the risks posed by complex mixtures of environmental contaminants, the design of more environmentally friendly pesticides and other biocides, the development of more sophisticated assays for the purposes of environmental risk assessment, and the improvement of procedures for statutory risk assessment.

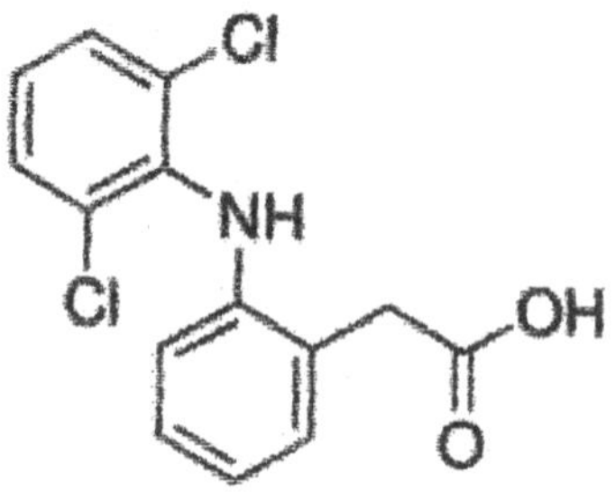

FIGURE 19.1 Structure of diclofenac.

DICLOFENAC AND VULTURES

Like some other veterinary medicines, diclofenac was originally marketed as a drug. Diclofenac is a nonsteroidal anti-inflamatory drug (NSAID) that is used extensively in human medicine (Figure 19.1). During the 1990s it came to be widely used as a painkiller for livestock in India. This practice was linked to the decline of three species of vulture in the Indian subcontinent (Green et al. 2006; Schultz and Charman 2004; Cuthbert et al. 2011). Three species of *Gyps* vulture, once common across the Indian subcontinent, have declined by more than 97% since 1992 (Senacha et al. 2008). These are the species *Gyps bengalensis*, *Gyps indicus*, and *Gyps tenuirostris.*

There are two aspects to this problem. First, diclofenac is fairly persistent in cattle, so it contaminates carrion eaten by the vultures. Second, it is quite toxic to vultures. When vultures were fed on meat from cattle that had died within two days of being dosed with the drug, 10% of them received a lethal dose in a single meal. Intoxication by diclofenac is associated with visceral gout, and many of the vultures found dead in Indian field studies had this condition. The use of diclofenac in this way was banned in India in 2006, but recent reports say that the ban has only been partially implemented (Cuthbert et al. 2011).

There are some parallels here with the effects of persistent organochlorine insecticides such as dieldrin upon predatory birds. Chlorine substitution in organic compounds tends to retard oxidative detoxication (Chapter 2), and thus increase their persistence in animals and birds. Second, we have another example of a toxic effect upon carnivorous species at the apex of a food pyramid that, as reported earlier (see Box 9.2), tend to be deficient in oxidative detoxifying enzymes. It would appear that there has not been the same evolutionary pressure to develop enzymes of this type in carnivores as has been the case with herbivores and omnivores. This metabolic deficiency is believed to make carnivores more efficient bioaccumulators of persistent lipophilic pollutants than the herbivores and omnivores below them in the food pyramid. It has been suggested that the high toxicity of this chlorinated drug to vultures is related to their poor ability to detoxify it by metabolism.

This decline of vultures has had consequences for humans as well as for the natural environment. The Zoroastrian community of India has traditionally left bodies of the dead in "towers of silence." In these, vultures and other scavengers have consumed human remains, thereby serving the cause of public hygiene. This practice is threatened by the severe declines that have been reported.

THE LARGE-SCALE DECLINE OF BEES AND OTHER POLLINATORS

In Chapter 12 there was discussion about the sublethal effects of pyrethroids and neonicotinoids, two groups of neurotoxic insecticides, upon bees and other pollinating insects (e.g., hover flies). There have been widespread declines of honeybees in recent years, and there has been some suggestion that the sublethal effects of these insecticides could be a contributory factor. The success of some agricultural and orchard crops is dependent upon pollination by insects, so we are dealing here with an economic issue as well as an ecological one.

Recent work has shown that sublethal doses of neonicotinoid insecticides can have behavioral effects upon honeybees at levels similar to those found in the field. These can impair the ability of bees to locate sources of nectar or to navigate their way back to the hive. The difficult questions to answer are: (1) To what extent are effects of this kind actually occurring in the field at the present time? (2) To what extent are such effects responsible for population declines of bees or failure of pollination in economic crops?

A complicating factor is the possibility that other pesticides currently in use may act as synergists of these insecticides in the field. It has already been shown that members of both types of insecticide can be synergized by other pesticides that inhibit oxidative detoxication by the cytochrome P450-based oxygenases. Prominent here are the ergosterol biosynthesis inhibitor (EBI) fungicides, such as prochloraz, imazalil, or triflumizole (see Walker 2008; Pilling et al. 1995; Schmuck et al. 2003). The risks of this happening are apparent if an insecticide is mixed with an EBI fungicide in the spray tank before application to a crop. They are less apparent if the insecticide and the EBI fungicide are applied independently of one another but come together in a plant. This is clearly a risk where both pesticides are systemic. For example, one component may be applied as an insecticide, and the other as a fungicide to the same crop, and these applications may come independently, at different times. Following systemic distribution, both compounds may be present in the same locations at the same time within the crop. So, for example, a bee may take in both compounds when it consumes nectar. The way is then open for the fungicide to potentiate (synergize) the toxicity of the insecticide within the bee if the quantities of residues are high enough (see Chapters 2 and 12).

Some systemic pesticides can persist in their active forms for considerable periods of time within the plant. Thus, a later application of a different pesticide may result in both it (e.g., an insecticide) and its synergist (e.g., a fungicide) coexisting in the vascular system or the nectar at the same time. Field studies are needed to establish whether synergistic effects of this kind pose risks to bees or other pollinators (e.g., hoverflies) under realistic conditions in the field. A number of research projects are underway that seek answers to these questions.

MARINE POLLUTION BY POLYISOBUTENE (PIB)

In 2013 there occurred two serious pollution incidents along the south coast of Devon and Cornwall in the United Kingdom. These were investigated by the Devon Wildlife Trust and reported in its bulletin "Protecting Wildlife for the Future" (2013). They resembled earlier cases of marine pollution by crude oil following the wreckage of oil tankers (Chapter 7), in that dead and dying seabirds were washed up that were contaminated with an oily residue. It was estimated that at least 4,000 seabirds were found along these shores, taking the two incidents together.

The species most affected were guillemots, with smaller numbers of fulmars, gannets, puffins, and herring gulls. They had become coated by the polymer polyisobutene (PIB), which, like crude oil, adheres tenaciously to feathers. The birds became immobilized, unable to fly or swim, and eventually died from starvation or hypothermia.

PIB is used in a wide range of manufacturing processes, including the production of cling film and adhesive tape. It is commonly used as a thickening agent in engine oils, and this was apparently the source of it in the present case. It is believed that the problem arose because ships had discharged it from within the English Channel. Moves are currently afoot to make this practice illegal.

EFFECTS OF MIXTURES OF ENVIRONMENTAL CHEMICALS

Interest in the possibility of synergistic interactions of pesticides in the field is one example of a wider concern about the possible ecological effects of mixtures of environmental contaminants. The ecotoxicity tests described here, which provide the basis for statutory environmental risk assessment, are usually only applied to individual chemicals. Only rarely are they performed using mixtures to discover whether there is potentiation of toxicity due to interactions between chemicals (Chapter 2).

In the first place, it is not feasible to test all possible combinations of pesticides and other industrial chemicals for possible enhancement of toxicity. This would be much too difficult and costly. On the other hand, such testing may be carried out where there is reasonable concern about the effects of certain mixtures that exist in the environment or are to be released into it. The possible interaction between pyrethroid or neonicotinoid insecticides with EBI fungicides, referred to in the previous section, is a case in point.

The discovery that ethinylestradiol (EE2), a component of contraceptive pills, can cause the feminization of male fish in the neighborhood of sewage outfalls (see Chapter 15) has stimulated interest in the possible environmental effects of pharmaceuticals more generally. What are now described as pharmaceuticals and personal care products (PPCPs) are known to be widely distributed in the aquatic environment, often in complex mixtures, albeit usually at very low concentrations. Questions have been asked about possible environmental effects of complex mixtures of them. One study of this kind investigated the possible effects of a mixture of beta blocker drugs (propranolol, metaprol, and nadolol) at environmentally realistic concentrations upon aquatic invertebrates (Huggett et al. 2002). In this case no significant effects were found, but many more combinations of environmental pharmaceuticals await investigation.

In approaching this problem it seems sensible to take account of the mechanisms by which potentiation of toxicity may occur (refer to Box 2.2) (Walker 2009). One very important example is the inhibition of the detoxication of one potential pollutant by another environmental chemical that is present in the same ecosystem. An understanding of mechanisms of this kind can ensure that the use of limited resources is focused upon those combinations of pollutants that are most likely to give rise to potentiation of toxicity.

THE DESIGN OF MORE ECOFRIENDLY PESTICIDES AND BIOCIDES

Many of the recognized problems of chemical pollution have involved pesticides or biocides, and examples of this were given in Section 2 of the present book. This is not surprising since the reason for producing such compounds is to cause harm to pests, vectors, and other organisms that present problems to the human race. In designing new, more environmentally friendly pesticides, the aim is to produce molecules that are both of limited environmental persistence and highly selective between target and nontarget organisms, but this is not easy to achieve.

Following recognition of the problems with persistent organochlorine insecticides such as DDT and dieldrin there was a move toward the production of less persistent pesticides that were not subject to significant biomagnification in food chains. In the first place, the earlier persistent compounds were extensively replaced by other existing less persistent ones. For example, certain carbamate or organophosphorous insecticides replaced persistent organochlorine ones for certain purposes. In the longer term effective new pyrethroid and neonicotinoid insecticides of limited persistence were introduced. So, in time, the old persistent compounds that had caused environmental problems were replaced by newer, more environmentally friendly ones.

Recent decades have also seen the emergence of new herbicides and fungicides, in replacement of older ones (e.g., organomercury fungicides) that were regarded as being environmentally hazardous. The EBI fungicides have generally been very successful, but not without problems. Some of them (e.g., procloraz) can strongly synergize the toxicity of pyrethroids and neonicotinoids (see discussion above and Chapter 12).

This is an ever-changing scenario. With time, pests, vectors, and weeds develop resistance to pesticides (Chapter 5). This is particularly a problem if pesticides are overused. A similar problem exists with the overuse of antibiotics, which has led to the evolution of resistant strains of microorganisms that are the cause of some major diseases (e.g., tuberculosis and malaria). If these trends continue, it will become impossible to control certain pests or human diseases with existing chemicals, and there will be a requirement for the development of new ones that overcome such problems.

There have also been recent concerns about possible ecological effects of pesticides, for example, sublethal neurotoxic effects of insecticides upon the behavior of bees (see above and Chapter 12). If these are substantiated, the quest will be on for more environmentally friendly alternatives.

The success of the rational design of new pesticides to control pests and diseases has depended on the exploitation of scientific discoveries in the broad field of biochemical toxicology and new technologies associated with them (Chapter 6). Future innovations will depend on continuing progress in relevant scientific fields and the development of related technologies. Hopefully more ecofriendly pesticides will be introduced in the course of time. We are now touching upon economic and political issues and priorities that lie outside of the scope of the present book.

IMPROVED METHODS OF ECOTOXICITY TESTING

In recent decades there has been growing interest in the development of more sophisticated methods of ecotoxicity testing to replace certain existing requirements of statutory environmental risk assessment, e.g., lethal toxicity tests to measure median lethal dose (LD_{50}) or median concentration (LC_{50}) (Chapter 2). Ecotoxicologists as well as environmentalists and animal welfare organizations have shared this interest (Walker 1998). A case has been made for the development of new *mechanistic biomarker assays* that could provide the basis for new in vitro tests. Such assays could also be valuable in field trials, e.g., of pesticides. In principle, they could be used to relate levels of pesticides found in exposed animals to the extent of measurable toxic effects that they are known to have at the biochemical or physiological levels. From an ecological point of view it should be possible to define threshold biomarker responses beyond which population declines may be expected to occur.

In recent decades there has been great progress in the fields of biochemistry and molecular biology. In medical biochemistry the rapid recent development of kits for the diagnosis of human diseases has shown what is possible with long-term planning and prudent investment. This new technology has the potential to facilitate the design of effective tests to identify toxic effects of chemicals upon free-living organisms—that is, effects at the physiological or biochemical level. But such progress does not come cheaply. Once again, we enter the fields of politics and economics, which lie outside the scope of the present text.

IMPROVEMENTS IN PROCEDURES FOR STATUTORY RISK ASSESSMENT OF ENVIRONMENTAL CHEMICALS

In Chapter 18 an overview was given of current legislation for risk assessment of environmental chemicals in developed countries, drawing attention to the fact that requirements for toxicity testing are usually more stringent for pesticides and biocides than they are for the general run of new industrial chemicals that are expected to be released into the environment (see also Chapter 2). The REACH legislation of the European Union was used as a particular example, drawing attention to certain aspects of it that have been subject to criticism (see also Walker 2006).

Statutory requirements for risk assessment of pesticides or biocides usually fall short of what ecotoxicologists would regard as true environmental risk assessment because they do not involve a proper assessment of risk at the level of population (for further discussion of this issue see Walker et al. 2012, Chapter 17). That said, there may occasionally be a requirement for a field trial if the initial risk assessment is inconclusive (see Chapters 2 and 18). If properly designed, such a field trial has the potential to estimate risk at the population level.

Sometimes, new evidence of harmful side effects of chemicals has brought changes in legislation. This happened, for example, when new restrictions or outright bans on persistent organochlorine insecticides and organomercury fungicides came into force during the second half of the twentieth century (Chapters 9 and 11). Much more recently, certain restrictions were placed on the use of neonicotinoid insecticides because of the perceived risk to bees.

It is to be hoped that current research will lead to improved procedures for ecotoxicity testing, including more sophisticated in vitro tests, and that some of these will, in due course, be incorporated into statutory requirements for risk assessment. There is, however, the danger that bureaucratic procedures become "cast in tablets of stone" and fail to keep pace with scientific progress. It is important that regulatory requirements remain flexible and responsive to new discoveries. It has been argued that there should be greater scientific involvement in the bureaucratic process in the future (Walker 2006).

SUMMARY

During the latter part of the twentieth century came recognition of harmful effects of chemicals upon the natural environment. To a large extent these effects were a consequence of human activity. Many of the examples given here are of the effects of pesticides and other man-made chemicals. In developed countries this discovery has led to the introduction of legally enforced bans and restrictions on the marketing and release of chemicals causing pollution. Persistent pesticides (e.g., organochlorine insecticides such as DDT and dieldrin) and some recalcitrant industrial chemicals (e.g., PCBs) were subject to these bans and were replaced with more biodegradable alternatives. In the course of time, the environmental levels of these chemicals fell, and there was some evidence of the recovery of natural populations that had been affected by them. Interest grew in the design of new, readily biodegradable and environmentally friendly pesticides.

While this was broadly true of developing countries, the situation was different in some developing countries where pesticides banned elsewhere continued to be used for the control of serious agricultural pests and vectors of disease—sometimes on a large scale. Some examples of pollution problems in developing countries are given in Chapter 17, although the picture is sketchy and incomplete because of the shortage of reliable information. International agencies such as the Food and Agriculture Organization (FAO) and World Health Organization (WHO) have been concerned about this issue, have investigated and reported on pollution problems in these countries, and continue to do so.

Into the twenty-first century some old pollution problems with highly persistent pollutants still remain, albeit to a much reduced extent. There are still areas with unacceptably high levels of persistent highly chlorinated pollutants and of methylmercury, for example. An ongoing problem is the emergence of new resistant strains of pest species and vectors of disease toward insecticides. This brings a need for the design of new pesticides that can overcome resistance problems.

Some new pollution problems have also emerged. One of the most dramatic has been the catastrophic decline of three species of vulture in India that has occurred since the early 1990s. This has been attributed to the use of diclofenac as a veterinary drug in cattle. This is a chlorinated drug that is highly toxic to these birds. There are certain parallels here with the organochlorine insecticides described in Chapter 9.

Into the twenty-first century there has been growing interest in sublethal effects of pollutants. It has become clear from ecotoxicological studies that sublethal effects as well as lethal ones can cause the decline of populations. The decline of birds of prey due to DDE-induced eggshell thinning and the decline of the dog whelk due to the action of TBT are cases in point. Currently, there is concern that the neurotoxic pyrethroid and neonicotinoid, which can affect the behavior of bees, may thereby have contributed to declines of bee populations. These are subtle issues, more difficult to investigate than mass poisoning incidents.

Hopefully recent rapid progress in biochemical toxicology will lead to improved methods for the investigation of pollution—and to the development of more refined methods of ecotoxicity testing to support environmental risk assessment. Not least, the development of new mechanistic biomarker assays that can be used in conjunction with population studies should lead to clearer insights into the effects of pollutants at the levels of population, community, and ecosystem.

FURTHER READING

Green, R.E., Taggart, M.A., Das, D., et al. 2006. Collapse of the Asian vulture populations: Risk of mortality from the veterinary drug diclofenac in carcasses of dead cattle. *Journal of Applied Ecology* 43: 949–956. Reports on the effect of diclofenac on vultures in India.

Glossary

2,4-D: 2,4-Dichlorophenoxyacetic acid. A widely used PGR herbicide. Used in domestic gardens as well as in agriculture.

2,4,5-T: 2,4,5-Trichlorophenoxyacetic acid. A plant growth regulator (PGR) herbicide used as a defoliant. Constituent of Agent Orange used in the Vietnam War (see Chapter 13). The formulation used in Vietnam contained dioxins as impurities, and thus constituted a human health hazard.

A esterase: An enzyme that degrades OPs that are in the oxon form. Birds are deficient in this enzyme in their blood, thus making them more susceptible than mammals to certain OPs (see Chapter 12).

Acaricide: A pesticide that controls mites (*Acarina*).

Acetylcholinesterase: An enzyme that degrades acetylcholine. Acetylcholine is a neurotransmitter that carries messages across cholinergic nerve synapses. Organophosphorous and carbamate insecticides act by inhibiting acetylcholinesterase of the nerve synapse. By so doing, they disrupt transmission of messages across synapses (see Chapter 10).

Acid rain: Rain that is made acidic (below pH 5.6) by the oxides of nitrogen and sulfur (see Chapter 7).

Aflatoxin B1: A toxin produced by the fungus *Aspergillus flavus* that is found on badly stored ground nuts. It is a liver carcinogen (hepatocarcinogen) that has sometimes had toxic effects upon poultry.

Agent Orange: A defoliant containing the herbicide 2,4,5-T. A formulation containing dioxins as a by-product was used during the Vietnam War. It constituted a human health risk.

Ah receptor: Aryl hydrocarbon receptor. A protein found in the cytosol of cells that binds aromatic hydrocarbons such as benzo(a)pyrene and other planar molecules, e.g., coplanar PCBs and dioxins.

Ah receptor-mediated toxicity: Toxic effects produced when certain planar compounds, e.g., coplanar PCBs or dioxins, bind to the Ah receptor.

Aldicarb: A systemic carbamate insecticide of high mammalian toxicity that is formulated as granules.

Aldrin: A chlorinated cyclodiene, once widely used as a soil insecticide. In biota, it is transformed rapidly to the much more stable and persistent insecticide dieldrin. It became subject to widespread bans because of its harmful effects (see Chapter 9).

Algae: Simple plants that contain the pigment chlorophyll and are able to carry out photosynthesis. Seaweeds such as bladder wrack are brown algae. Green algae form algal blooms during the process of eutrophication (see Chapter 7).

Alkaline: Having a pH above 7 (see Box 7.1).

Ames test: A test used to identify mutagens (see Chapter 2).

Amino acid: An acid containing an amino group, e.g., glycine, alanine, tyrosine, leucine, methionine, etc. They are the building units for proteins.

Ammonia: A gas; formula NH_3. It is formed when organic residues decompose. It dissolves in water to form ammonium hydroxide, which is alkaline.

Ammonium ion: NH_4^+. A constituent of ammonium salts such as ammonium chloride (NH_4Cl).

Androgen: A male sex hormone such as testosterone. They are all steroids, and regulate the development of male secondary sexual characteristics.

***Anser* geese:** Grey geese, e.g., grey lag goose and pink-footed goose. Migratory geese that were poisoned by the OP carbophenothion in the UK in 1972–1973 (see Chapter 10).

Anticoagulant rodenticides: Rodenticides that act by inhibiting the formation of clotting proteins and so causing hemorrhaging (see Chapter 16).

Antifouling paint: Paint applied to boats that prevents the attachment of limpets and other marine organisms to the hulls of boats. Tributyl tin has been used as an active ingredient in such paints (see Chapter 11).

Antiknocks: Compounds, such as tetraalkyl lead, that have been used as petrol additives to prevent "knock" (semiexplosive burning) in internal combustion engines.

Aroclor mixtures: Mixtures of PCBs (see Chapter 13).

***Aspergillus* spp:** Fungi belonging to this group synthesize aflatoxin.

ATP: Adenosine triphosphate, a biochemical molecule with a key role in energy transformation.

Atropine: Acts as an antagonist of acetylcholine on some cholinergic junctions of the nervous system. Has been used as an antidote in cases of organophosphorous poisoning (see Chapter 10).

Avermectins: Natural products that have been used as helminthicides, i.e., to destroy parasitic worms.

B esterases: Enzymes that hydrolyze esters and are inhibited by OPs.

***Bacillus thuringiensis* (B.T.):** A bacterium that produces an insecticidal toxin. Preparations of it have been used as insecticides. The gene that encodes for B.T. toxin has been incorporated into some genetically manipulated organisms, e.g., transgenic cotton.

Batrachotoxin: A toxin produced by certain frogs (e.g. *Dendrobates* spp.). It has been used as an arrow poison by Amerindians in South America.

Benzo(a)pyrene: A polycyclicaromatic hydrocarbon formed during the incomplete combustion of organic compounds. It is a carcinogen, found in tobacco smoke and tar.

Beta blockers: Extensively used drugs that act upon beta adrenergic receptors (receptors on certain nerve synapses where noradrenaline acts as a neurotransmitter). They are antagonists of noradrenaline.

Bioaccumulation: Animals ingest pollutants present in their food. In the course of time the concentration of residue increases in the tissues of the animal. The bioaccumulation factor is given by the ratio of the concentration of pollutant in the tissues of the animal to the concentration of the pollutant in its food.

Biochemical toxicology: The biochemical basis of toxicology, especially mode of action and metabolism.

Bioconcentration: Relating to the uptake of chemicals by animals from their ambient medium, by aquatic organisms from surrounding water and by terrestrial animals from air. Bioconcentration factor is given by the ratio of the concentration in the organism to the concentration in the ambient medium.

Biodegradable: Relating to environmental chemicals; their degradation by living organisms predominantly by enzymic action.

Biodiversity: A measure of the variety of living organisms on earth, including genetic diversity. In ecology the stability of ecosystems depends on it. In agriculture, monocultures tend to be unstable systems; problems with pests, for example, may be due to the disappearance of their natural enemies during the course of intensive farming. The misuse of pesticides can reduce biological diversity.

Biological oxygen demand (BOD): In aquatic pollution, the quantity of oxygen required in water to oxidize organic residues such as raw sewage.

Biomagnification: With reference to food chains, the ratio of the concentration of pollutant in one trophic level to the concentration of the same pollutant in a lower trophic level of the same ecosystem.

Biomarker: A biological response to a chemical at the level of the individual organism or below that demonstrates a departure from normal status. A *mechanistic* biomarker provides a measure of toxic effect. Some biomarkers only provide a measure of exposure. Biomarker responses provide the basis for biomarker assays (see Chapter 2).

Bipyridyl herbicide: A herbicide such as diquat or paraquat that contains two linked pyridine rings. Pyridine is a six-membered ring that contains one atom of nitrogen.

Bisphenol A: An industrial chemical that is an endocrine disruptor.

Bordeaux mixture: A copper precipitate that has long been used as a fungicide, e.g., on vines and to control potato blight.

Brodifacoum: An anticoagulant rodenticide.

BTO: British Trust for Ornithology.

Buffer solution: A solution with reserve acidity or alkalinity. When acids or bases are added to buffer solutions, pH change occurs relatively slowly. Surface waters with high buffering capacity are less affected by acid rain than are other waters that lack this characteristic.

Calcium ATP-ase: An enzyme system that drives the uptake of calcium into the shell gland of birds. When p,p′-DDE causes eggshell thinning there is evidence that inhibition of this enzyme system is the cause. However, there is still uncertainty whether the pollutant acts directly on calcium ATP-ase.

Calux assay: A bioassay for detecting the presence of coplanar PCBs, dioxins, and other pollutants that cause Ah receptor-mediated toxicity.

Carbamate insecticides: Insecticides that are derivatives of carbamic acid, e.g., carbaryl and carbofuran. Like OPs, they act as inhibitors of acetylcholinesterase.

Carbaryl: A carbamate insecticide.

Carbendazim: An EBI fungicide.

Carbofuran: A carbamate insecticide.

Carcinogen: A chemical that causes cancer.

Carrying capacity: In ecology, the capacity of a defined area (habitat) to support particular species. Capacity, in this case, refers to population density.

CFCs: Volatile chlorofluorocarbon compounds that have been used as refrigerants and propellants. They have been implicated in the thinning of the ozone layer (see Chapter 7).

Chemical warfare: In the present text this term is taken in its widest sense—to include plant-animal warfare, which has been a factor in the evolution of life on earth, as well as the use of chemical weapons by humans for hunting, tribal warfare, and control of pests and diseases.

Chloracne: A skin condition caused by environmental chemicals, including dioxins.

Chlordane: An organochlorine insecticide.

Chlorfenvinphos: An OP.

***Chrysanthemum* spp:** Members of the Compositae, some of which have been a source of the natural insecticide pyrethrum.

Clone: A population of organisms produced asexually, e.g., by aphids.

Clothianidin: A neonicotinoid insecticide.

Common Agricultural Policy (CAP): A policy followed by the European Union that has included paying subsidies to farmers to adopt ecofriendly practices such as "set aside" and conserved areas that are not cultivated or treated with pesticides.

Congener: A member of a group of chemicals that share common structural features, e.g., PCBs.

Contaminant: 1. An environmental chemical that is produced by man and does not occur naturally. 2. Sometimes also applied to a naturally occurring chemical when it occurs at an abnormally high level. This may be due to the activities of man or to extreme natural events such as volcanic eruption, earthquake, fall of meteors, etc.

Coplanar: In the same plane—as with coplanar PCBs.

Corvids: Birds that are members of the crow family.

Coumarol: A naturally occurring chemical found in some grasses that can cause hemorrhaging in animals that consume it.

Curare: A naturally occurring chemical that has been used as an arrow poison.

Cuticle: In insects, the hard surface layer of the body.

Cyclodiene insecticides: A group of chlorinated insecticides that include aldrin, dieldrin, and heptachlor.

Cytochrome P450: An oxidizing enzyme that exists in many forms and is very important in the metabolism of organic pollutants by animals.

DDT: A once widely used organochlorine insecticide, now much less used due to bans placed by many countries.

Detergent: Synthetic surfactant. There are anionic, cationic, and non-ionic detergents.

Diazinon: An OP once widely used for controlling agricultural pests and for dipping sheep to control parasites.

Diclofenac: A human medicine that has been used as a veterinary medicine in the Indian subcontinent. The latter practice has led to the dramatic decline of vultures (see Chapter 19).

Dieldrin: A chlorinated cyclodiene insecticide that has been largely banned. It has caused declines of predatory birds in Western Europe.

Dimethoate: A systemic OP.

Dioxin equivalent: A measure of Ah receptor-mediated toxicity caused by mixtures of dioxins and PCBs.

Dioxins: Industrial by-products. Highly persistent organochlorine compounds that can cause Ah receptor-mediated toxicity.

Diphenacoum: A second-generation anticoagulant rodenticide that is effective against some warfarin-resistant rodents.

Disyston: A systemic OP.

Dose-response curve: In toxicity testing, a graph that relates the dose of a chemical to the toxic response. Used to estimate values for LD_{50}, LC_{50}, etc.

Dove, turtle (*Streptopelia purpur*): A once common species on UK farmland that declined sharply during the latter part of the twentieth century. Intensive arable farming, including the heavy use of herbicides, may have been the primary cause.

Eagle, bald (*Haliaeetus leucocephalus*): A large raptor affected by residues of DDT in North America (see Chapter 9).

Ecosystem: A collection of populations that occur in the same place and at the same time that can interact with each other and their physical and chemical surroundings.

Ecotoxicity: The estimated toxicity of a chemical to free-living organisms that inhabit the natural environment. Such estimates are made, very largely, using surrogate species (see Chapter 2).

EE2: Ethinylestradiol. A potent estrogen that is widely used in oral contraceptives.

Emulsifiable concentrate: A type of formulation often used for insoluble pesticides (see Box 6.1).

Endocrine disruption: A hormonal imbalance initiated by exposure to a chemical that leads to alterations in development, growth, and reproduction in organisms and their progeny (see Chapter 15).

Endocrine glands: Glands such as the thyroid, pituitary, and adrenal that secrete hormones.

Endoplasmic reticulum: A membranous network found in many cell types—notably the liver of vertebrates. Lipophilic pollutants readily diffuse into it, after which they are metabolized into more water-soluble products by enzymes—especially forms of cytochrome P450 (see Box 2.1).

Environmental risk assessment: In principle, estimation of the risks presented by environmental chemicals to natural populations. In practice, this is difficult and costly to accomplish, and current statutory requirements for environmental risk assessment tend to fall short of this objective (see Chapters 18 and 19).

Enzymes: Proteins that catalyze a wide variety of chemical reactions within living organisms.

Ergosterol biosynthesis inhibitors (EBIs): A group of fungicides, many of them systemic, which act by retarding the synthesis of ergosterol. They do this by inhibiting forms of cytochrome P450 that are involved in ergosterol biosynthesis. They can also act as synergists for other pesticides that are detoxified by this type of enzyme (see Box 2.2 and Chapter 19).

Ester: An organic salt, formed by the combination of an organic acid with an organic base. If an ester is hydrolyzed, an organic acid and an organic base are released.

Estrogens: Hormones such as estradiol that promote the development of female sexual characteristics.

Eutrophication: Excessive enrichment of surface waters by nitrates and other nutrients that leads to oxygen deficiency and the appearance of algal blooms (see Chapter 7).

Flycatcher, spotted (*Musciapa striata*): An insectivorous bird whose decline in agricultural areas has been related to intensive arable farming and the use of herbicides.

Food and Agriculture Organization of the United Nations (FAO): An international organization that has been concerned with the misuse of pesticides in the developing world, and has published reports on this subject.

Formulation: Pesticides are often marketed as formulations that are ready to be applied as they are (e.g., dusts) or mixed with water (wettable powders or emulsifiable concentrates). Formulations include additives such as emulsifiers and stabilizers (see Box 6.1).

Fungicide: A pesticide designed to control fungal diseases, especially on agricultural and horticultural crops.

GABA receptor: A receptor found in both vertebrate and invertebrate nervous systems. It represents the site of action of the cyclodiene insecticides dieldrin and endrin. It also represents the site of action for aldrin because this compound is rapidly converted to dieldrin within animals—and it is dieldrin that blocks the receptor (see Chapter 9).

Gaia theory: A global model originally proposed by James Lovelock that addresses pollution problems such as global warming due to greenhouse gases and damage to the ozone layer caused by CFCs (see Chapter 7).

Gannet (*Sula bassanus*): A large piscivorous seabird of the Atlantic that was shown to be affected by p,p′-DDE residues in Canada (see Chapter 9).

Gas chromatography: An analytical technique that has been widely used to detect low levels of pollutants in biota.

Gland: Organs of animals that secrete enzymes, hormones, etc. Endocrine glands secrete hormones such as testosterone and estradiol. Pollutants that cause endocrine disruption are discussed in Chapter 15.

Glyphosate: A widely used herbicide that can achieve total weed control. It has the advantage of becoming inactive when it reaches soil, principally because it becomes tightly bound to soil minerals.

Goshawk (*Accipiter gentilis*): A large hawk that was affected by methylmercury seed dressings in Scandinavia (see Chapter 11).

Grebe, western (*Aechmophorus occidentalis*): A piscivorous bird that was poisoned by the organochlorine insecticide DDD (rhothane) on Clear Lake,

California. One of the first examples of lethal effects on a predator following biomagnification in a food chain (see Chapter 9).

Greenhouse gases: Certain atmospheric gases, especially carbon dioxide, that can cause global warming (see Chapter 7).

Half-life: The time that it takes for the concentration of a chemical to be halved. Half-lives are measured for the decay of radioisotopes and the loss of pollutants and other chemicals from animals, plants, and soils (biological half-lives) (see Box 9.1).

Heptachlor: A chlorinated cyclodiene related to aldrin. In animals it is quickly metabolized to heptachlor epoxide, a persistent compound related to dieldrin. Dieldrin and heptachlor epoxide both act upon GABA receptors (see Chapter 9).

Homeostasis: The maintenance of a constant state within living organisms with regard to such parameters as pH of tissue fluids, body temperature, salt concentrations, etc.

Hydrolysis: The breakdown of chemicals such as esters by the action of water.

Hydrophobic: Water hating. Hydrophobic compounds tend to have low polarity and low water solubility. They tend to be lipophilic (lipid loving). Within living organisms there are hydrophobic zones such as fat depots, lipoproteins, and membranous structures within cells, e.g., the endoplasmic reticulum. Lipophilic pollutants, e.g., persistent organochlorine insecticides and PCBs, tend to partition into these from the adjacent aqueous environment.

Imazalil: An EBI fungicide that has been widely used.

Imidacloprid: A neonicotinoid insecticide.

Immunotoxicity: Toxic effects on the immune system.

Imposex: The masculinization of certain female molluscs such as the dog whelk, caused, for example, by certain organotin compounds. The development of a penis leads to the females becoming infertile (see Chapter 11).

Insecticide: A pesticide that is toxic to insects.

International Atomic Energy Agency (IAEA): An international organization based in Vienna, Austria, that has certain responsibilities as a watch dog ensuring that nuclear weapons are not developed in contradiction of international treaties. They also have been involved in investigations of the possible illegal development of chemical weapons, e.g., by Iraq when led by Saddam Hussein, and the misuse of pesticides in the African continent.

Invertebrates: Animals without backbones.

Kdr: Knockdown resistance. A resistance mechanism reported in houseflies and some other insects to DDT and pyrethroids. This cross-resistance between two different types of insecticide is due to their sharing the same mode of action. A mutant form of a sodium channel that is insensitive to both these insecticides confers resistance (see Chapter 5).

Kestrel, American (*Falco sparverius*): Toxicological studies with this species demonstrated that the DDT metabolite p,p′-DDE can cause eggshell thinning (Wiemeyer and Porter 1970) (see Chapter 9).

Kestrel, common (*Falco tinnunculus*): A species that was affected by dieldrin in Great Britain (see Chapter 9).

Ketoconazole: An EBI fungicide.

LC_{50}: In ecotoxicity testing, a median lethal concentration (see Chapter 2).

LD_{50}: In ecotoxicity testing, a median lethal dose (see Chapter 2).

Lead tetraalkyl: Once widely used as an antiknock in petrol. Now subject to extensive bans. Restrictions placed because of human health hazard when used in this way. Has also caused serious poisoning incidents involving wading birds (see Chapter 11).

Limonite: Brown coal. It has a high sulfur content and when burned forms significant quantities of the gas sulfur dioxide. There was serious damage caused to pine forests of Silesia, in Czechoslovakia and Poland, during the communist era because of the burning of brown coal.

Lipophilic: Lipid loving. Lipophilic compounds have low polarity, high fat solubility, and low water solubility. They tend to be stored in the fat depots of animals.

Locust, desert (*Schistocerca gregaria*): A serious crop pest found from North Africa to India. Like the desert locust, it has been subject to control by aerial spraying (see definition below).

Locust, migratory (*Schistocerca migratoria*): A serious crop pest found throughout Africa, South Eurasia, Australia, and New Zealand. It has been controlled by aerial spraying with insecticides, sometimes causing extensive damage to fauna (see Chapter 17).

Loon, common (*Gavia immer*): A piscivorous diving bird found on the lakes of North America that has been closely studied with respect to long-standing organomercury pollution (see Chapter 11).

Malathion: An OP with relatively low mammalian toxicity.

Mass screening: In the quest for new pesticides, candidate compounds have sometimes been tested against a range of organisms seeking biological activity (see Chapter 6).

MCPA: Methoxychlorophenoxycetic acid. A plant growth regulator herbicide.

Mechanistic biomarkers: Biomarkers such as acetylcholinesterase inhibition that measure toxic effects.

Mercury: Both organic and inorganic forms of mercury are important as pollutants. Organomercury compounds are described in Chapter 11.

Mesocosm: Model ecosystems such as ponds that are used to study the effects of pollutants (see Chapter 4).

Metabolism: Chemical change catalyzed by enzymes.

Metabolite: A product of metabolism. Fat-soluble organic pollutants are converted by enzymes into polar metabolites to facilitate excretion.

Metals: Typically metals are elements that form cations (positively charged ions) and conduct electricity. They form salts. Some metals are regarded as pollutants when present in the environment at relatively high concentrations, e.g., at mine workings. Examples include lead, copper, cadmium, nickel, and mercury.

Methiocarb: A carbamate that is used as both an insecticide and an acaricide.

Methylarsenic compounds: Both naturally occurring and man-made (see Chapter 11).

Methylmercury compounds: Organometallic compounds that are both man-made and naturally occurring. Methylmercury fungicides were once widely used. Methylmercury compounds have caused pollution problems (see Chapter 11).

Microtox assay: A bioassay that has been used to detect pollutants (see Chapter 2).

Mite, red spider: A pest species that has appeared in orchards as the result of overuse of insecticides. Heavy use of insecticides has sometimes greatly reduced or wiped out capsid bugs, which are natural enemies of red spider mites.

Molecular biology: The study of the molecular basis of life.

Molluscs: A large class of invertebrates. Many are marine organisms, e.g., shellfish and squids. Some inhabit freshwater, e.g., snails and bivalves; some are terrestrial, e.g., slugs and snails.

Mosquito, malarial: Some *Anopheles* species are vectors for the malarial parasite. DDT and pyrethroids have been used on a large scale to control these mosquitoes. Large swampy areas have sometimes been aerially sprayed with insecticides causing severe damage to wildlife. Strains of malaria mosquitoes that are resistant to insecticides have emerged, making it more difficult to control the disease (see Chapter 17).

Mutagen: A chemical that causes mutations of DNA. Benzo(a)pyrene, a constituent of tobacco smoke, is a potent mutagen.

Mutant: A mutated form of DNA, a protein, or an organism. Resistant strains of pests that emerge when pesticides are overused often possess aberrant forms of the site of action (e.g., the enzyme acetylcholinesterase). Such aberrant forms are often termed mutant forms (see Chapter 5).

Mycotoxin: A fungal toxin.

Nanoparticles: Small particles that are found in environmental samples. Of particular interest are engineered nanoparticles (see Chapter 17).

Neonicotinoid insecticides: A group of insecticides whose structure and mode of action resemble that of nicotine. Imidacloprid is an example (see Chapter 12).

Nerve gases: OP compounds, designed to be used as chemical weapons for use against humans. Like the related insecticides, they are acetylcholinesterase inhibitors. Examples include soman and sarin.

Neuropathy target esterase (NTE): An esterase of the nervous system. When inhibited by certain OPs, delayed neuropathy may follow (see Chapter 10).

Neurotoxicity: Toxic effects of chemicals upon the nervous system of animals.

Nicotine: A naturally occurring compound found in the tobacco plant that has been used as an insecticide.

Nicotinic receptor: An acetylcholine receptor found in both vertebrate and invertebrate nervous systems that is sensitive to nicotine and neonicotinoid insecticides.

Nitrate ion: An ion found in soils and present in inorganic nitrate fertilizers. It can leach from soils into surface waters. High levels of nitrate in surface waters can lead to eutrophication and the appearance of algal blooms (see Chapters 4 and 7).

Nitrogen oxides: Gases that contribute to acid rain (see Chapter 7).

Organic compounds: The compounds of carbon.

Organoarsenic compounds: Methylarsenic compounds are both manufactured (e.g., pesticides) and naturally occurring (cf. methylmercury compounds) (see Chapter 11).

Organochlorine insecticides: A once widely used group of insecticides. Examples include DDT, aldrin, dieldrin, and benzehexachloride (BHC). The more persistent ones caused environmental problems associated with biomagnification in food chains, and became extensively banned (see Chapter 9).

Organohalogen compounds: Organic compounds contain halogen elements (chlorine, fluorine, iodine, or bromine).

Organolead: Tetraethyl lead was once widely used as an additive to petrol. Now much restricted in use because of human health hazards (see Chapter 11).

Organomercury: Methylmercury occurs naturally and has been produced commercially, e.g., in the manufacture of fungicides. Methylmercury fungicides are extensively banned because of environmental problems.

Organophosphorous insecticides (OPs): A once widely used group of neurotoxic pesticides that act as cholinesterase inhibitors. Some are highly toxic to mammals (see Chapter 10).

Organotin: Tributyl tin has been used as a component of antifouling paint, and has caused endocrine disruption in marine molluscs such as the dog whelk. It is now subject to restrictions in its use (see Chapter 11).

Owl, barn: A rodent feeder that has sometimes been poisoned by superwarfarins (see Chapter 16).

Owl, screech: A rodent feeder that has sometimes been poisoned by superwarfarins (see Chapter 16).

Oxon: With reference to OPs, forms (often metabolites) that have a double bond between phosphorous and oxygen (see Chapter 10).

Oxyradicals: Highly reactive forms of oxygen that cause cellular damage in animals and plants. They can be generated by oxidative enzymes. The toxicity of bipyridyl herbicides such as diquat and paraquat is related to the generation of oxyradicals.

p,p′-DDD: Metabolite of DDT formed under anaerobic conditions. Also active ingredient of insecticide rhothane (see Chapter 9).

p,p′-DDE: Highly persistent metabolite of the insecticide DDT that is subject to strong biomagnifications in both terrestrial and aquatic food chains. Has been implicated in the thinning of avian eggshells.

Paraquat: A bipyridyl herbicide that is used as a total weed killer. It becomes inactivated in soils due to strong binding to clay minerals.

Partridge, grey: A farmland bird that has undergone widespread decline in Western Europe. This has been attributed to a shortage of insect food for chicks in intensively farmed areas. The shortage of insects has been related to the shortage of the weeds that they feed upon (see Chapter 14).

PCBs: Polychlorinated biphenyls (see Chapter 13).

Peach-potato aphid (*Myzus persicae*): A serious pest of cereal crops that has developed resistance to insecticides. Clones that have developed resistance to OPs possess multiple copies of a gene that encodes for an esterase (see Chapters 5 and 10).

Pelican, brown (*Pelicanus occidentalis*): A piscivorous bird that produced thin eggshells as a consequence of exposure to p,p′-DDE (see Chapter 9).

Peregrine (*Falco peregrinus*): A predator that feeds upon other birds. It was affected by organochlorine insecticides in Western Europe and North America. In Great Britain there was a severe decline attributed to the effects of cyclodiene insecticides, particularly dieldrin. In North America declines also occurred, but they were related to eggshell thinning caused by p,p′-DDE rather than the lethal or sublethal effects of cyclodienes (see Chapter 9).

Persistent organic pollutants (POPs): A collection of organic pollutants that are very persistent in the environment that have been identified by the United Nations Environment Program (UNEP) and termed the dirty dozen. Persistent organochlorine insecticides, PCBs, and dioxins are represented here (see Chapter 13).

Phloem: The part of the vascular system of plants that is involved in the transport of sugars and other nutrients. Some systemic pesticides are transported by the phloem.

Physostigmine: A naturally occurring carbamate that has neurotoxicity. It has been used as a truth drug by African tribes and served as a model for carbamate insecticides (see Chapter 10).

Pigeon, wood (*Columba palumbus*): A vertebrate pest in Western Europe, causing damage to vegetable crops. Many individuals were poisoned when diedrin was used as a cereal seed dressing in the middle of the twentieth century (see Chapter 9).

Piscivorous: Feeding upon fish, e.g., birds such as cormorant, gannet, or puffin.

Plant growth regulator (PGR) herbicides: Widely used herbicides such as MCPA, 2,4-D, mecoprop, etc., that disrupt the processes of growth in plants (see Chapter 14).

Plant toxins: Compounds produced by plants that are toxic to animals. They are thought to have evolved to protect plants against grazing by animals (see Chapter 3).

Pollutant: See definition given in Chapter 1.

Polycyclic aromatic hydrocarbons (PAHs): Compounds that consist of fused aromatic rings. Some of them, e.g., benzo(a)pyrene, are carcinogens (see Chapter 7).

Polyhalogenated aromatic hydrocarbons (PHAHs): PCBs, dioxins, and related compounds (see Chapter 13).

Polyisobutene (PIB): A synthetic polymer that has recently been recognized as a marine pollutant (see Chapter 19).

Population dynamics: Changes in population density (see Chapter 4).

Postsynaptic membrane: With reference to nerve junctions (synapses), when the presynaptic membrane of one nerve receives a message, it releases a neurotransmitter (e.g., acetylcholine) that crosses the synapse to interact with a receptor located on the *postsynaptic membrane* of a second nerve.

The message is then carried along the second nerve. The whole process is termed synaptic transmission. For examples, see Chapter 9.

Potentiation: With reference to the toxicity of mixtures, potentiation occurs when the toxicity of a mixture substantially exceeds the summation of the toxicities of its individual components.

Prochloraz: An EBI fungicide.

Pyrethroid insecticides: A group of synthetic insecticides that bear a structural resemblance to natural pyrethrins.

Pyrethrum: A natural product used as an insecticide (see Chapter 12).

QSAR: Quantitative structure-activity relationship. In principle, the correlation of molecular structure with biological activity (e.g., toxicity). Following this principle, a molecule that has the shape and distribution of electrical charge that permit it to bind tightly to a receptor for a neurotransmitter is likely to act as a nerve poison. QSAR models have been developed to predict the toxicity of environmental chemicals such as pesticides. The approach has attracted much interest because it has, in the long term, the potential to reduce the requirement for standard toxicity tests using animals, during the process of risk assessment (see Chapters 2 and 18).

Quelea quelea **(red-billed quelea):** A bird that is a serious pest of crops in sub-Saharan Africa and has been controlled by the aerial spraying of pesticides.

Radioisotope: An isomer of an element that emits radiation (see Chapter 7).

Raptor: Bird of prey.

Raven (*Corvus corax*): A large member of the crow family. Poisoned by anticoagulant rodenticide during rat control programs (see Chapter 16).

Resistance: With regard to pesticides, the appearance of resistant strains as a consequence of the selective pressure of pesticides (see Chapter 5).

Ricin: A toxin found in the castor oil plant (*Ricinis communis*). Among the most poisonous chemicals known to mammals, including humans.

RIVPACS: River Invertebrate Prediction and Classification System (see Chapter 4).

Rodenticide: A pesticide that controls rats, mice, and other rodents.

Rotenone: A naturally occurring insecticide found in the plant *Derris elliptica*.

RSPB: Royal Society for the Protection of Birds (Great Britain).

Scientific Committee on Problems of the Environment (SCOPE): An international organization that organizes meetings and generates publications on problems of pollution.

Selective toxicity: Differences between species, strains, sexes, and age groups regarding their susceptibility to the toxic effects of a chemical.

Shell gland: In birds, the gland within which eggs are formed. Calcium ions are imported into it and used for the formation of eggshells. Eggshell thinning caused by p,p′-DDE is characterized by a failure in the transport of blood calcium into the eggshell gland.

Site of action: A structure within an organism with which a chemical interacts to cause a toxic effect, for example, a receptor for a neurotransmitter or the active site of an enzyme.

Society of Environmental Toxicology and Chemistry (SETAC): An international society that organizes meetings and publishes a journal dealing with problems of pollution.

Sodium channels: Channels through biological that allow the movement of sodium ions. The operation of such channels in nerve membranes mediates the passage of electrical impulses along nerves. Some insecticides, e.g., DDT and pyrethroids, can bind to sites on these channels and so cause disruption of the transmission of nerve impulses.

Sparrowhawk, European (*Accipiter nisus*): A bird-eating raptor that suffered declines in Great Britain and some other Western European countries as a consequence of exposure to dieldrin and other cyclodiene insecticides (see Chapter 9).

Spatfall: Production of young oysters.

Steric hindrance: The hindrance of chemical reactions caused by the presence of chemical structures or groups that "get in the way."

Strychnine: A plant toxin that has been used to control moles.

Sulfur: An element that has been used as a fungicide.

Sulfur dioxide: A gas that is formed when sulfur or sulfides are burned. It has been a major factor in the formation of acid rain (see Chapter 7).

Superwarfarins: Potent anticoagulant rodenticides that are successors to warfarin (see Chapter 16).

Surfactant: A chemical (e.g., a soap or detergent) that has surface activity.

Surrogate species: With reference to toxicity testing, a species that is used as an alternative to another for testing purposes. For example, laboratory rats are often used as surrogates for other mammals when testing pesticides as part of statutory risk assessment (see Chapter 18).

Synapse: Junction between two nerves—the ending of one and the beginning of another. A neurotransmitter such as acetylcholine is released at a nerve ending and then carries a message to another nerve across the synapse.

Synergism: With reference to the toxicity of mixtures, a situation where one compound (the synergist) increases the toxicity of another in the same mixture. The synergist itself expresses no toxicity at the dose at which it is applied.

Systemic: With reference to pesticides, when applied to plants, systemic chemicals are transported by the vascular system to reach untreated organs or tissues.

Tern, Caspian (*Hydroprogne caspia*): A piscivorous bird found in North America that was adversely affected by PCBs and dioxins (see Chapter 13).

Tetraalkyl lead: An organometallic compound once extensively used as an additive to petrol (see Chapter 11).

Tetrodotoxin: A toxin found in puffer fish.

Thiacloprid: A neonicotinoid insecticide that can be synergized by EBIs.

Tobacco bud worm (*Heliothis virescens*): A serious pest of cotton in the southern United States. Strains that are highly resistant to pyrethroid insecticides have emerged (see Chapters 5 and 12).

Triazine herbicides: Soil-acting herbicides. Atrazine has sometimes reached unacceptably high levels in surface waters and drinking water.

Trophic level: Level in the food chain. Predators exist in the highest trophic levels. For examples, see Chapter 9.

Tsetse fly (*Glossina* spp.): A vector for trypanosomes (agents causing sleeping sickness) found over large areas of Africa. Extensive spraying operations, including aerial spraying, have raised concerns about pollution (see Chapter 17).

Urea herbicide: Derivatives of urea used as soil-acting herbicides.

Vascular system of plants: In higher plants, there is a well-developed vascular system comprising the xylem and phloem. Systemic pesticides can move around plants in the vascular system.

Vitellogenin: In fish, a protein whose synthesis is stimulated by estrogens. This response has been used as the basis of a biomarker assay of exposure.

Vultures: Carrion feeding birds that have experienced dramatic population declines in India as a consequence of ingestion of diclofenal (see Chapter 19).

Wettable powders: Formulations of solid pesticides that can be dispersed in water and applied as sprays.

Whelk, dog: A marine mollusc that has suffered declines due to pollution by tributyl tin (see Chapter 11).

World Health Organization (WHO): An international organization that has been involved in monitoring large-scale use of pesticides for vector control.

Xenobiotic: Literally, a foreign compound. A compound that is not part of the normal biochemistry of the organism in question. It is important to remember that a compound that is foreign to one organism may be normal to another. Man-made compounds are usually foreign to wild animals.

Xylem: Part of the vascular system of higher plants. Water and nutrients taken up from the soil move upward in it to reach the aerial parts of the plants (i.e., leaves, flowers, fruit, etc.). This upward transport is referred to as the transpiration stream.

References

Agosta, G.W. [1996] *Bombardier Beetles and Fever Trees: a Close up Look at Chemical Warfare and Signals in Animals and Plants*. Reading MA: Addison Wesley.

Ahman, B. and Ahman, G. [1994] Radiocaesium in Swedish reindeer after the Chernobyl fall-out. Seasonal variations and long term decline. *Health Physics* 66 503–512.

Alzieu, C. [2000] Impact of tributyl tin on marine invertebrates. *Ecotoxicology* 9 71–76.

Andrews, J.E., Brimblecombe, P., Jickells, T.D. et al [1996] *An Introduction to Environmental Chemistry*. Oxford, UK: Blackwell Sciences.

Ashton, F.M, Crafts, A.S. [1973] *Mode of Action of Herbicides*. New York: John Wiley.

Bailey, S., Bunyan, P.J., Jennings, D.M. et al [1974] Hazards to wildlife from the use of DDT in orchards II A further study. *Agro-Ecosystems* 1 323–338.

Ballantyne, B. and Marrs, T.C. [1992] *Clinical and Experimental Toxicology of Organophosphates and Carbamates*. Oxford, UK: Butterworth/Heinemann.

Basu, N., Klevanic, K., Gamberg, M. et al [2005] Effects of mercury in neurochemical receptor binding characteristics in wild mink. *Environmental Toxicology and Chemistry* 24 1444–1450.

Beauvais, S.L., Jones, S.B., Brewer, S.K. et al [2000] Physiological measures of neurotoxicity of diazinon and malathion to rainbow trout and their correlation with behavioural measures. *Environmental Toxicology and Chemistry* 19 1875–1880.

Begon, M., Mortimer, M. and Thompson, D.J.A. [1996] *Population Ecology: a unified study of animals and plants*, 3rd Ed. Oxford: Blackwell Scientific.

Berg, W., Johnels, A., Sjostrand, B., et al [1966] Mercury content of feathers of Swedish birds over the past 100 years. *Oikos* 17 71–83.

Boon, J.P., Ejgenram, F., Everarts, J.M. et al [1989] A structure-activity relationship-approach to metabolism of PCBs in marine animals from different trophic levels. *Marine Environmental Research* 27 159–176.

Borg, K., Erne, K., Hanko, E. et al [1970] Experimental secondary methyl mercury poisoning in the goshawk. *Environmental Pollution* 1 91–104.

Brealey, C.J., Walker, C.H. and Baldwin, B.C. [1980] 'A' esterase activities in relation to the differential toxicity of pirimiphos methyl *Pesticide Science* 11 546–554.

Briggs, S.A. [1992] Basic Guide to Pesticides Their Characteristics and Hazards. Taylor and Francis.

British Trust for Ornithology. [2011] Annual Report.

Broley, C.L. [1958] *Plight of the American Bald Eagle Audubon Magazine* 60 162–171.

Brooks, G.T. [1966] *World Review of Pest Control* 5 62–84.

Brooks [1969] Investigations with some biodegradable dieldrin analogues Proceedings 5th Insecticide and Fungicide Conference. Brighton, UK 472–477.

Brooks G.T [1974] Chlorinated Insecticides Volume 2 Biological and Environmental Aspects CRC Press USA: Cleveland.

Brooks, G.T., Harrison, A. and Lewis, S.E. [1970] *Biochemical Pharmacology* 19 255–273.

Bruggers, R.L. and Elliott, C.C.H. [1989] *Quelea Quelea: Africas bird pest* Oxford, UK: Oxford University Press.

Bryce-Smith, D. [1971] Lead pollution from petrol Chemistry in Britain 7 284–286.

Bull, K.R., Every, W.J., Freestone, P. et al [1983] Alkyl lead pollution and bird poisoning on the Mersey Estuary, UK 1979–1981 Environmental Pollution [Series A] 31 239–259.

Calow, P. [Ed] [1998] *Handbook of Ecotoxicology*. Oxford UK: Blackwell Science.

Campbell, L.H., Avery, M.I, Donald, P. et al [1997] A Review of the Indirect Effects of Pesticides on Birds JNCC Report No. 227 Joint Nature Conservation Committee, Peterborough, U.K.

Caquet, T., Lagadic, L. and Sheffield, S.R. [2000] Mesocosms in Ecotoxicology. *Reviews in Environmental Contamination and Toxicology* 84 71–79.

Carson, R. [1962] Silent Spring. Boston, Houghton Mifflin.

Chanin, P.R.F. and Jefferies, D.J. [1978] The decline of the otter in Britain: Analysis of hunting records and discussion of causes *Biological Journal of the Linnaean Society* 10 305–328.

Chipman, J.K. and Walker, C.H. [1979] The metabolism of dieldrin and two of its analogues: the relationship between rates of microsomal metabolism and rates of excretion of metabolites in the male rat. *Biochemical Pharmacology* 28 1337–1445.

Colin, M.E. and Belzunces, L.P. [1992] Evidence of synergy between prochloraz and deltamethrin; a convenient biological approach *Pesticide Science* 36 115–119.

Combes, R.B. Dandrea, J. and Balls, M. [2003] Registration, Authorisation and Restriction of Chemicals [REACH] Proposal FRAME and the Royal Commission on Environmental Pollution common recommendations for assessing risks posed by chemicalsunder the EC REACH system ATLA [Altern. Lab. Anim.] 31 529–535.

Copping, L.G. and Duke, S.O. [2007] Natural Products Used as Commercially as Crop Protection Agents. *Pest Management Science* 63 524–554.

Cramp. S, Bourne, W.R.P., and Sanders, D. [1976] The Seabirds of Britain and Ireland 3rd Edition. London: Collins.

Cresswell, J.E. [2011] Meta analysis of experiments testing the effects of a neonicotinoid [imidacloprid] on honeybees. *Ecotoxicology* 20 149–157.

Crick, H.Q.P., Baillie, S.R., Balmer, D.E. et al [1998] Breeding Birds in the Wider Countryside: Their Conservation Status UK Thetford Research Report 198, British Trust for Ornithology.

Crosby, D.G. [1998] *Environmental Toxicology and Chemistry*. New York: Oxford University Press.

Cuthbert, R.T., Taggart, M.A., Prakash, V. et al [2011] Effectiveness of action in India to reduce exposure of Gyps vultures to the toxic veterinary drug diclofenac PlosOne,6,e19069.

Darwin, Charles [1859] The Origin of Species. London: John Murray.

Dawkins, C.R [1976] The Selfish Gene. Oxford, UK: Oxford University Press.

Desbrow, C., Routledge, E.J. Brighty, G.C. et al [1998] Identification of estrogenic chemicals in STW effluent. 1. Chemical fractionation and in vitro biological screening. *Environmental Science and Technology* 32 1549–1558.

Devon Wildlife Trust [2013] Wild Devon Magazine Summer 2013 p18 www.devonwildlifetrust.org

Devonshire, A.L. [1991] Role of esterases in the resistance of insects. IN Biochemical Society Transactions, 19 755–759.

Devonshire, A.L., Byrne, F.J., Moores, G.D. et al [1998] Biochemical and molecular characterisation of insecticide insensitive cholinesterase in resistant insects IN Doctor, B.P. et al [Eds] Structure and Function of Cholinesterases and Related Proteins. New York: Plenum Press. 491–496.

Douthwaite, R.J. [1992] Effects of DDT on the fish eagle population of Lake Kariba in Zimbabwe. IBIS 134 250–258.

Eason, C.T. and Spurr, E.B. [1995] Review of toxicity and impacts of brodifacoum to non-target birds and mammals in New Zealand. *New Zealand Journal of Zoology* 22 371–379.

Eason, C.T., Murphy, E.C., Wright, G.R.G. et al [2002] Assessment of risks of brodifacoum to non target birds and mammals in *New Zealand Ecotoxicology* 11 35–48.

Edwards, C.A. [1976] *Persistent Pesticides in the Environment* 2nd Edition. London: CRC Press.

Edwards, T. [1994] Chernobyl National Geographic 186 100–115.

Ehrlich , P.R. and Raven, P.H. [1964] Butterflies and Plants: a study in coevolution. *Evolution* 18 586–608.

Eldefrawi, M.E. and Eldefrawi, A.T. [1990] Nervous System Based Insecticides IN Safer Insecticides-Development and Use. E. Hodgson and R.J. Kuhr [Eds]. New York: Marcel Dekker.

Elliott, J.E., Norstrom, R.J. and Keith, J.A. [1988] Organochlorines and eggshell thinning in Northern gannets *Environmental Pollution* 52 81–102.

Elliott, M. [1977] Synthetic Pyrethroids. American Chemical Society Symposium Series 42 1–28.

Encyclopedia Britannica 2013 Plastic pollution. Written by Charles Moore. http://www.EBchecked /topic/1589019/plastic pollution.

Environmental Health Criteria No. 86 [1989] Mercury: Environmental Aspects. Geneva: WHO.

Environmental Health Criteria No. 91 [1989]. Aldrin and Dieldrin. Geneva: WHO.

Environmental Health Criteria No. 101 [1990] Methyl Mercury. Geneva: WHO.

Evers, D.C., Savoy, L.J., De Sorbo, C.R. et al [2008] Adverse effects from environmental mercury loads on breeding common loons *Ecotoxicology* 17 69–81.

Fergusson, D. [1994] The effects of 4-hydroxy coumarin anticoagulants on birds and the development of techniques for non-destructive monitoring their effects. PhD Thesis University of Reading, UK.

Fest, C. and Schmidt, K.-J. [1982] Chemistry of Organophosphorous Compounds. Berlin: Springer Verlag.

French-Constant, T.A., Rochelau, G., Streichen, J.C. et al [1993] A point mutation in a Drosophila GABA receptor that confers insecticide resistance. *Nature* 363 149.

Finney, D.J. [1964] Probit Analysis 2nd Edition. Cambridge UK: Cambridge University Press.

Flannigan, B. [1991] Mycotoxins IN D'Mello, J.P.F. et al [Eds], Toxic Substances in Crop Plants. London: Royal Society of Chemistry 225–250.

Fuchs, P. [1967] Death of birds caused by application of seed dressings in the Netherlands. Mededel Rijksfaculteit Landbouwweetenschappen Gent 32 855–859.

Garrison P.M., Tullis, K. and Aarts, J.M.JG.G. [1996] Species specific recombinant cell lines as bioassay systems for the detection of dioxin-like chemicals. *Fundamental and Applied Toxicology* 30 , 194–203.

Gilbertson, M., Fox, G.A., and Bowerman, W.W. (Eds) [1998] Trends in Levels and Effects of Persistent Toxic Substances in the Great Lakes. Dordrecht, Kluwer Academic Publishers.

Green, R.E., Taggart, M.A., Das, D. et al [2006] Collapse of the Asian vulture populations: risk of mortality from the veterinary drug diclofenac in carcasses of dead cattle. *Journal of Applied Ecology* 43 949–956.

Greig-Smith, P., Frampton, G. and Hardy, A.R. [1992] Pesticides, Cereal Farming and the Environment London, HMSO.

Grue, C.E., Hart, A.D.M. and Mineau, P. [1991] Biological consequences of depressed cholinesterase activity in wildlife IN Mineau, P. [Ed] Cholinestease inhibiting insecticides–their Impact on Wildlife and the Environment 151–210 Amsterdam: Elsevier.

Guillette, L.J., Pickford D.B., Crain, D.A. et al [1996]. Reduction in penis size and plasma testosterone levels in juvenile alligators in a contaminated environment. *General and Comparative Entomology* 101 32–42.

Hamilton, G.A., Hunter, K., Ritchie, A.S. et al [1976] Poisoning of wild geese by carbophenothion-treated winter wheat *Pesticide Science* 7 175–183.

Harborne, J.B. [1993] Introduction to Ecological Biochemistry 4th Edition. London: Academic Press.

Hardy, A.R. [1990] Estimating exposure: The identification of species at risk and routes of exposure In. L. Somerville, and C.H. Walker [Eds] Pesticide Effects on Terrestrial Wildlife 81–98 London: Taylor and Francis.

Harris, Robert [2003] Pompeii. BCA publication. Random House Group Ltd., Polmont, Stirlingshire, UK.

Harrison, R. and Lunt, G.G. [1980] Biological Membranes 2nd Edition. Glasgow: Blackie.

Hassall, K.A. [1990] The Biochemistry and uses of Pesticides Basingstoke, Hants Macmillan.

Hegdal, P.L. and Colvin, B.A. [1988] Potential hazard to Eastern Screech owls and other raptors of brodifacoum bait used for vole control in orchards. *Environmental Toxicology and Chemistry* 7 245–260.

Henry, M., Beguin, M., Requier, F. et al [2012] A common pesticide decreases survival and foraging success in honey bees *Science Express Report* March 29 2012 pp 1–4.

Hodgson, E. and Guthrie, F.E. [1980] Introduction to Biochemical Toxicology. New York: Elsevier.

Hodgson, E. and Kuhr, R.J. [1990] Safer Insecticides-Development and Use. New York: Marcel Dekker.

Hooper, M.J. et al [1989]. Organophosphate exposure in hawks inhabiting orchards during winter dormant spraying. *Bulletin of Environmental Contamination and Toxicology* 42 651–660.

Hopkin, S.P. [1989] Ecophysiology of metals in terrestrial invertebrates. Barking, UK: Elsevier Science.

House, W.A., Leach, D., Long, J.L.A. et al [1997] Microorganic compounds in the Humber rivers. *The Science of the Total Environment* 194–195 [Special Issue] 357–372.

Howald, G.R., Mineau, P., Elliot, J.E. et al [1999] Brodifacoum poisonong of avian scavengers during rat control at a seabird colony. *Ecotoxicology* 8 431–437.

Howells, G. [1995] Acid Rain and Acid Waters 2nd Edition. Hemel Hempstead, Ellis Horwood.

Huckle, K.R., Warburton, P.H., Forbes, S. et al Studies on the fate of flocoumafen in the Japanese quail Xenobiotica 19 51–62.

Huggett, D.B., Brooks, B.W., Peterson, B. et al [2002]. Toxicity of select beta blocker pharmaceuticals on aquatic organisms. *Archives of Environmental Contamination and Toxicology* 41 229–235.

Hunt, E.G. and Bischoff, A.I. [1960] Inimical effects on wildlife of periodic DDD applications to Clear Lake, California. *Fish and Game* 46 91–106.

Ilyinskikh, E.N., Ilyinskikh, N.N. and Ilyinskikh, I.N. [1999] IN Peakall, D.B., Walker, C.H. and Migula, P. [Eds] Biomarkers: a Pragmatic Basis for Remediation of Severe Pollution in Eastern Europe. NATO Science Series "Environmental Security" Vol. 54 Dordrecht Kluwer Academic Publishers.

International Atomic Energy Agency (IAEA) [1997] Report IAEA–TECDOC–931 Organochlorine Insecticides in African Ecosystems Report of a Final Research Coordination Meeting IAEA and FAO.

Iwasa, T., Motoyama, N., Ambrose, J.T. et al [2004] Mechanism of differential toxicity of neonicotinois insecticides in the honeybee. *Crop Protection* 23 371–378.

Jensen, S. [1966] Report of a new chemical hazard *New Scientist* 32 612.

Jeschke, P. and Nauen, R. [2008] Neonicotinoids from zero to hero *Pest Management Science* 64 1084–1098.

Johnson, M.K. [1992] Molecular events in delayed neuropathy: experimental aspects of neuropathy target esterase IN Ballantyne and Marrs 1992.

Jones, D.M., Bennett, D. and Elgar, K.E. [1978] Deaths of owls traced to insecticide-treated timber London, *Nature*, 272 52.

Karr, J.R. [1981] Assessment of Biotic Integrity using fish communities *Fisheries* 6 21–27.

Kidd, K.A., Blanchfield, P.J., Mollis, K.H. et al [2007] Collapse of a fish population after exposure to a synthetic estrogen. Proceedings of the National Academy of the United States of America 104 8897–8901.

Koeman, J. and Pennings, J.H. [1970] An orientational survey on the side effects and environmental distribution of dieldrin in a tse–tse control in S.W. Kenya *Bulletin of Environmental Contamination and Toxicology* 5 164–170.

Koeman, J.H., Den Boer, W.M.J. Feith, H.H. et al [1978] Three years observation on side effects of helicopter applications of insecticides used to exterminate Glossina species in Nigeria. *Environmental Pollution* 15 31–59.

Koeman, J.H., Van Velsen-Blad, de Vries, R. et al [1973] Effects of PCB and DDE in Cormorants and evaluation of PCB residues in an experimental study. *Journal of Reproductive Physiology* [supplement] 19 353–364.

Korsloot, A., van Gestel, A.M. and van Straalen, N.M. [2004] Environmental Stress and Cellular Response in Arthropods. Boca Raton: CRC Press.

Kuhr, R.J. and Dorough, H.W. [1977] Carbamate Insecticides. Boca Raton, FL: CRC Press.

Leahy J.P. (Ed.) [1985] The Pyrethroid Insecticides. London: Taylor and Francis.

Lewis, D.F.V. [1996] Cytochrome P 450s Structure, Function and Mechanism. London: Taylor and Francis.

Livingstone, D.R., Moore, M.N. and Widdows, J. [1988]. Ecotoxicology; Biological effects measurement on molluscs and their use in impact assessment IN W. Salomans, B.L., Baynes, E.K. Duursma et al. Pollution of the North Sea: an Assessment. Berlin: Springer Verlag 624–637.

Lovelock, J. [1988] The Ages of Gaia. Oxford, UK: Oxford University Press.

Lovelock, J. [1982] Gaia: a New Look at Life on Earth. Oxford UK: Oxford University Press.

Lutgens, F.K. and Tarbuck, G.K. [1992] The Atmosphere: an Introduction to Meteorology. Hemel Hempstead, Prentice-Hall.

Lynch, J.M. and Wiseman, A. [1998] Environmental Biomonitoring: the Biotechnology Ecotoxicology Interface. Cambridge University Press.

Lyr, H. (Ed) [1987] Modern Selective Fungicides. Harlow, UK: Longman Scientific and Technical.

Mackness, M.I., Thompson, H.M. and Walker, C.H. [1987] Distinction between 'A' esterases and aryl esterases and implications for esterase classification. *Biochemical Journal* 245 293–296.

Macnair, M.R. [1987] Heavy Metal Tolerance in Plants: a Model Evolutionary System Trends in Evolution and Ecology 2 354–359.

Markussen, M.D., Heiberg, A.C., Fredholm, M. et al [2008] Differential expression of Cytochrome P 450 genes between bromodiolone-resistant and anticoagulant susceptible Norway rats. A possible role for pharmacokinetics *Pest Management Science* 64 239–248.

Maron, D.M. and Ames, B.N. [1983] Revised methods for the Salmonella mutagenicity test. *Mutation Research* 113 173–215.

Marrs, T.C., Maynard, R.L. and Sidell, F.R. [2007] Chemical Warfare Agents-Toxicology and Treatment 2nd Edition Chichester, UK: John Wiley and sons.

Matthews, G.A. [1979] Pesticide Application Methods London, Longmans.

Matthiessen, P., Sheahan, D., Harrison, R. et al [1995] Use of a Gammarus pulex assay to measure the effects of transient carbofuran run off from farmland. *Ecootoxicology and Environmental Safety* 30 111–119.

McCaffery, A.R. [1998] Resistance to Insecticides in Heliothine Lepidoptera: a global view. Philosophical Transactions of the Royal Society of London B 353 1735–1750.

Mellanby, K. [1967] Pesticides and Pollution. London: Collins.

Mineau, P. and Palmer, C. [2013] Neonicotinoid insecticides and birds. American Bird Conservancy http://www.abcbirds.org/abcprograms/policy/Neonic_FINAL.pdf

Moore, N.W. (Ed) [1966]. Pesticides in the environment and their effects upon wildlife. *Journal of Applied Ecology* 3 [Supplement].

Moore, N.M. and Walker, C.H. [1964] Organic chlorine insecticide residues in wild birds. *Nature* 201 1072–1073.

Morcillo, Y., Janer, G., O'Hara, S.C.M. et al [2004] Interaction of tributyl tin with cytochrome P450 and UDP glucuronyl transferase in fish: in vitro studies. *Environmental Toxicology and Chemistry* 23 990–996.

Moriarty, F. [1999] Ecotoxicology 3rd Edition. London: Academic Press.

Moriarty, F., ed. [1975] Organochlorine Insecticides: Persistent Organic Pollutants. London: Academic Press.

Mullie, W.C., Verwey, P.J., Berends, A.G. et al [1991] The impact of pesticides on palearctic migratory birds in the Western Sahel. ICBP Technical Publications No. 12 1991.

Nacci, D.E, Coiro, L. , Champlin, D. et al [2002] Predicting the occurrence of genetic adaptation to dioxin-like compounds in populations of the estuarine fish *Fundulus heteroclitus Environmental Toxicology and Chemistry* 21 1525–1532.

Nash, J.P., Kime, D.E. van der Ven et al [2004] Long term exposure to environmental concentrations of the pharmaceutical ethynyl estradiol causes reproductive failure in fish. *Environmental Health Perspectives* 112 1725–1733.

Natural Environment Research Council 1971 The sea bird wreck in the Irish Sea Autumn 1969 NERC Publications Series C No 4 1971.

Newman, M.C. [2010] Fundamentals of Ecotoxicology 3rd Edition. Boca Raton: CRC Press.

Newton, I. [1986] The Sparrowhawk Calton T. and A.D. Poyser.

Newton, I. and Wyllie, I. [1992] Recovery of a sparrowhawk population in relation to declining pesticide contamination. *Journal of Applied Ecology* 29 476–484.

Newton, I., Meek, E. and Little, B. [1978] Breeding Ecology of the Merlin in Northumberland British Birds 71 376–398.

Newton, I., Wyllie, I. and Freestone, P. [1990] Rodenticides in British Barn owls *Environmental Pollution* 68 101–117.

Nosengo, N. [2005] Fertilised to death. *Nature* 425 894–895.

Oberdorster, G. [2001] Pulmonary effects of inhaled ultrafine particles. International Archives of Occupational and Environmental Health. 74 1–8.

Odum, E.P. [1971] Fundamentals of Ecology. Philadelphia: W.B Saunders.

Osborn, Andrew [2011] How Chelyabinsk became synonymous with pollution. *Daily Telegraph* July 27th 2011.

Parker, P.J.-A.N. and Callaghan [1997] Esterase activity and allele frequency in field populations of Simulium Equinum exposed to organophosphate pollution. *Environmental Toxicology and Chemistry* 16 2550–2555.

Peakall, D.B., Walker, C.H. and Migula, P. [1997]. Biomarkers: a Pragmatic Basis for Remediation of Severe Pollution in Eastern Europe Dordrecht Kluwer: Academic Publishers.

Peakall, D.B. [1992] Animal Biomarkers as Pollution Indicators. London: Chapman and Hall.

Peakall, D.B. [1993] DDE-induced eggshell thinning: an environmental detective story Environmental Reviews 1 13–20.

Peakall, D.B. and Bart, J.R. [1983] Impacts of aerial application of insecticides on forest birds *Critical Reviews in Environmental Control* 13 117–165.

Peakall, D.B. and Shugart, L.R., Eds, [1993] Biomarkers: Research and Application in the field of Environmental Health. Berlin: Springer Verlag.

Persoone, G., Janssen, C. and De Coen, W. [2000] New Microbiotests for Routine Toxicity Screening and Biomonitoring. Kluwer Academic/Plenum.

Pesonen, M., Goksoyr, A., and Andersson, T. [1992] Expression of Cytochrome p450 in a primary culture of rainbow trout microsomes exposed to B Naphthoflavoneor 2,3,7,8-TCDD *Archives of Biochemistry and Biophysics* 292 228–233.

Pilling, E.D., Bromley-Challenor and Walker [1995] Mechanism of synergism between lamda cyhalothrin and prochloraz in the honeybee *Pesticide Biochemistry and Physiology* 51 1–11.

Potts, G.R. [2000] The Grey Partridge In D. Pain and J. Dixon [Eds] Bird Conservation and farming Policy in the European Union. London: Academic Press.

Potts, G.R. [1986] The Partridge. London: Collins.

Puinean, A.M., Elias, J., Slater, R. et al [2013] Development of a high throughput assay for the detection of the R81T mutation inthe the nicotinic acetylcholine receptor of neonicotinoid-resistant Myzus persicae *Pest Management Science* 69 195–199.

Purdom, C.E., Hardiman, P.A., Bye, V.V.J. et al [1994] Estrogenic effects of effluents from sewage treatment works *Chemistry and Ecology* 8 275–285.

Ramade, F. [1992] Precis d'ecotoxicology. Paris: Masson.

Ratcliffe, D.A. [1967] Decrease in Eggshell Weight in Certain Birds of Prey Nature 215 208–210.

Ratcliffe, D.A. [1993] The Peregrine Falcon Calton T. and A.D. Poyser.

Richards, P., Johnson, M., Ray, D. and Walker, C.H. [1999] Novel protein targets for organophophorous compounds. *Chemico-Biological Interactions* 119–120 503–512.

Robinson, J., Richardson, A. and Crabtree A.N. et al [1967] Organochlorine residues in marine organisms *Nature* 214 1307–1311.

Rodgers-Gray, T.P., Jobling, S., Kelly, C. et al [2001] Exposure of juvenile roach to treated sewage effluent induces dose-dependent and persistent disruption in gonadal duct development *Environmental Science and Technology* 35 462–470.

Salgado, V.L. [1999] Resistant target sites and insecticide discovery IN Brooks, G.T. and Roberts, T.R. [Eds] Pesticide Chemistry and Bioscience–the Food Environment Challenge *Cambridge Royal Society of Chemistry* pp 236–246.

Schmuck, R. [1999] No causal relation between gaucho seed dressing in sunflowers and the French bee syndrome Pflanzenschutz Nachrichten Bayer 52 257–299.

Schmuck, Stadtler, T. and Schmitt, H.W. [2003] Field reference of a synergistic effect observed in the laboratory between an EBI fungicide and a cloronicethyl insecticide in the honeybee *Pest Management Science* 59 279–286.

Schultz, Bara, S. and Charman, S. [2004] Diclofenac poisoning is widespread in declining vulture populations across the Indian sub-continent. Series B Biological Sciences. Proceedings of the Royal Society of London. Series B Biological Sciences Supplement 458–460.

Scott, J.G., Liu, N. and Zhimou, W. in Forms and Functions of Cytochrome P 450 Stegemann, J.J. and Livingstone D.R. [1998] *Comparative Biochemistry and Physiology* 121 C 147–155.

Scown, T.M., Van Aele, R. and Tyler, C.R. [2010] Do engineered nanoparticles present a threat in the aquatic environment? *Critical Reviews in Toxicology* 10. 653–670.

Senacha, K.R. Taggart, M.A., Rahmani, A.R. et al [2008] Diclofenac levels in livestock carcasses in India before the ban. *Journal of the Bombay Natural History Society* 105 148–161.

Sheahan, D.A., Brighty, G.C., Daniel, M. et al [2002] Estrogenic activity measured in a sewage treatmentworks treating industrial inputs containing high concentrations of alkylphenolic compounds—a case study. *Environmental Toxicology and Chemistry* 21 507–514.

Sheffield, S.R., Sawickaa-Kapustka, K., Cohen, J.B. et al IN R.F. Shore and B.A. Rattner (Eds) Ecotoxicology of Wild Mammals. Chichester, UK: John Wiley. 215–314.

Shore, R.F. and Rattner, B.A. (Eds) [2001] Ecotoxicology of Wild Mammals. John Wiley: Chichester UK.

Sibly, R.M., Newton, I. and Walker, C.H. [2000] Effects of dieldrin on population growth rate in the UK sparrowhawk *Journal of Applied Ecology* 7 540–546.

Somerville, L. and Walker, C.H. [1990] Pesticide Effects in Terrestrial Wildlife London: Taylor and Francis.

Somerville, M. and Greaves, M.P. [1987] Pesticide effects upon Soil Microflora. London: Taylor and Francis.

Sparks, T.C., Graves, J.B. and Leonard, B.R. [1993] Insecticide resistance and the tobacco bud worm: past, present and future. *Reviews in Pesticide Toxicology* 2 149–183.

Suett, D.L. [1986] Accelerated degradation of carbofuran in previously treated soil in the United Kingdom Crop Protection 5 165–169.

Sumpter, J.P. and Jopling, S. [1995] Vitellogenesis as a biomarker for estrogenic contamination of the marine environment. *Environmental Health Perspectives* 103 173–178.

Sussman, J.L., Harel, M. and Frolow, F. et al [1991] Atomic Structure of acetylcholinesterase from *Torpedo californica* a prototypic acetylcholine-binding protein. *Science* 253 872–879.

Thijssen, H.H.W. [1995] Warfarin-based rodenticides;mode of action and mechanism of resistance *Pesticide Science* 43 73–78.

Thompson, H.M. [2003] Behavioural effects of pesticides on bees their potential for use in risk assessment *Ecotoxicology* 12 317–330.

Timbrell, J. [1995] Introduction to Toxicology 2nd ED. London: Taylor and Francis.

Timbrell, J. [1999] Principles of Biochemical Toxicology 3rd edition London: Taylor and Francis.

Truhaut, R. [1977] Ecotoxicology: objectives, principles and perspectives. *Ecotoxicology and Environmental Safety* 1 151–173.

United Nations Environment Programme [UNEP] November 2011 'Bridging the Gap' A UNEP Synthesis Report from a meeting in Nairobi, Kenya ISBN 978–92–807–3229–0.

Van Emden, H.F. and Rothschild, M. (Eds) [2004] Insect and Bird Interactions. Andover, UK: Intercept publishers.

Van Straalen, N.M. [2004] Ecotoxicology becomes stress. Environmental Science and Technology, September 1 324–330.

Walker, C.H. [1975] Environmental Pollution by Chemicals 2nd Ed London Hutchinson.

Walker, C.H. [1978] Species differences in microsomal monooxygenase activities and their relation to biological half lives *Drug Metabolism Reviews* 7 (2) 295–323.

Walker, C.H. [1980] Species variations in some hepatic microsomal enzymes that metabolise xenobiotics. *Progress in Drug metabolism* 5 118–164.

Walker, C.H. [1983] Pesticides and birds; mechanisms of selective toxicity. *Agriculture, Environment and Ecosystems*. 9 211–216.

Walker, C.H. [1998] Alternative approaches and Tests in ecotoxicology: a review of the present position and the prospects for change taking into account ECVAM's duties, topic selection and test criteria ATLA [Alter. Lab. Anim.] 26 649–677].

Walker, C.H. [2004] Organochlorine insecticides and raptors in Great Britain IN Van Emden, H.F. and Rothschild, M. [Eds] Insect and Bird Interactions Intercept Ltd. UK: Andover 133–148.

Walker [2006] Ecotoxicity Testing of chemicals with particular reference to pesticides. *Pest Management Science* 62 571–583.

Walker, C.H. [2009] Organic Pollutants: an Ecotoxicological Perspective 2nd Edition. Boca Raton, FL: Taylor and Francis.

Walker, C.H. and El Zorgani, G.A. [1974] The comparative metabolism of HCE, a biodegradable analogue of dieldrin, by vertebrate species. *Archives of Environmental Contamination and Toxicology* 2 97–116.

Walker, C.H. and Livingstone, D.R. Eds [1992]. Persistent Pollutants in the Marine Environment. Special Publication of SETAC. Oxford, UK: Pergamon Press.

Walker, C.H. and Newton, I. [1999] Effects of cyclodienes on raptors in Britain–correction and updating of an earlier paper. *Ecotoxicology* 8 425–430.

Walker, C.H., Sibly, R.M., Hopkin, S.P. and Peakall, D.B. [2012] Principles of Ecotoxicology 4th edition Boca Raton, Fl: Taylor and Francis/CRC.

Weseloh, D.V., Teeple, S.M. and Gilbertson, M. [1983] double-crested cormorants of the great lakes: egg-laying parameters, reproductive failure and contaminant residues in eggs Lake Huron 1972–1973 *Canadian Jounal of Zoology* 61 427–436.

Whitehorn, P.R., O'Connor, S. Wackers, F.L. et al [2012] Neonicotinoid pesticide reduces bumble bee colony growth and queen production Science Express Report 29 March 2012 pp 1–3.

Wiemeyer, S.N. and Porter, R.D. [1970] DDE thins eggshells in captive American kestrels. *Nature* 227 737–738.

Williams, R.J., Brooke, D.N. Clare, R.W. et al [1996] Rosemount Pesticide Transport Studies 1987–1993 Report No. 129 Wallingford, UK: Institute of Hydrology [NERC].

Wolfe, M.F., Atkeson, T., Bowerman, W. et al [2007] Wildlife indicators. IN R. Harris, et al. [Eds] Ecosystem response to Mercury Contamination: Indicators of Change. CRC Press; SETAC 1232–189.

Wolfe, M.F., Schwarzbach, S. and Sulaiman, R.A. [1998] The effects of mercury on wildlife: a comprehensive review. *Environmental Toxicology and Chemistry* 17 146–160.

Wood, M. [1995] Environmental Soil Biology 2nd Ed. Glasgow, Blackie.

Wright, J.F. [1995] Development and use of a system for predicting the macroinvertebrate fauna found in flowing water. *Australian Journal of Ecology* 20 181–197.

Index

Q

R